Materials Processing in
High Gravity

Materials Processing in High Gravity

Edited by

Liya L. Regel and
William R. Wilcox

Springer Science+Business Media, LLC

Library of Congress Cataloging-in-Publication Data

On file

Proceedings of the Second International Workshop on Materials Processing in High Gravity, held June 6–12, 1993, in Potsdam, New York

ISBN 978-1-4613-6073-5   ISBN 978-1-4615-2520-2 (eBook)
DOI 10.1007/978-1-4615-2520-2

© 1994 Springer Science+Business Media New York
Originally published by Plenum Press, New York in 1994

PREFACE

There are two motives for studying materials processing in centrifuges. First, such research improves our understanding of the influence of acceleration and convection on materials processing. Second, there are commercial opportunities for production of unique and improved materials that cannot be prepared under normal earth conditions or in space.

Through a combination of experiments and theory, we are gaining an understanding of centrifugation on phenomena of importance to materials processing. We find that it is necessary to consider not only acceleration, but also the Coriolis effect and the variation of acceleration with position. As one consequence, the vigor of buoyancy-driven convection is sometimes increased by centrifugation and sometimes decreased. Similarly, the tendency of the convection to become unstable or oscillatory may either be increased or decreased by centrifugation. On the other hand, the observed effects of centrifugation on product quality have largely gone unexplained.

This volume constitutes the proceedings of The Second International Workshop on Materials Processing at High Gravity, hosted by Clarkson University in June of 1993. The concept for a workshop on materials processing in centrifuges was born at a series of informal meetings held in Paris in 1990. The First International Workshop on Materials Processing at High Gravity was held in May of 1991 in Dubna, USSR, on the banks of the Volga River. The proceedings of this workshop was published in 1992 as a special issue of the Journal of Crystal Growth. The Second International Workshop on Materials Processing at High Gravity was planned for Potsdam, NY, on the banks of the Racquette River. Like Dubna, the site of the first workshop, Potsdam is a village in a scenic area, concentrating on technology, with many foreign visitors, and a sizable river running through it. During this workshop, we dedicated Clarkson's unique centrifuge facility for materials processing research and related flow visualization.

The Second International Workshop on Materials Processing at High Gravity was supported by a grant from the United States National Science Foundation.

In this book, the symbol "g" is sometimes used to represent Earth's gravity and other times, especially in equations and dimensionless numbers, g designates the total acceleration vector. When g represents Earth's gravity, then the magnitude of the total acceleration is expressed by Ng, where N is any positive number.

Liya L. Regel and William R. Wilcox

International Center for
Gravity Materials Science and Applications
Potsdam, NY

CONTENTS

INTRODUCTION TO MATERIALS PROCESSING IN LARGE CENTRIFUGES

Liya L. Regel and William R. Wilcox

International Center for Gravity Materials Science and Applications
Clarkson University
Potsdam, NY 13699-5700, USA

ABSTRACT

This volume represents the proceedings of The Second International Workshop on Materials Processing at High Gravity, hosted by Clarkson University in June of 1993. Evidence continues to demonstrate the unique and advantageous features of centrifugation during materials processing.

In this book, the symbol "g" is sometimes used to represent Earth's gravity and other times, especially in equations and dimensionless numbers, g designates the total acceleration vector. When g represents Earth's gravity, then the magnitude of the total acceleration is expressed by Ng, where N is any positive number.

Through a combination of experiments and theory, we are gaining an understanding of centrifugation on phenomena of importance to materials processing. We find that it is necessary to consider not only acceleration, but also the Coriolis effect and the variation of acceleration with position. As one consequence, the vigor of buoyancy-driven convection is sometimes increased by centrifugation and sometimes decreased. Similarly, the tendency of the convection to become unstable or oscillatory may either be increased or decreased by centrifugation. On the other hand, the observed effects of centrifugation on product quality have largely gone unexplained. In this introduction, we summarize our current understanding of centrifugation effects as gained from the Workshop and the papers in this volume. We conclude with recommendations for future research efforts.

WORKSHOPS

The concept for a workshop on materials processing in centrifuges was born at a series of meetings held in Paris in 1990. The purpose of these meetings was to try to interpret the surprising results of Huguette Rodot and Liya Regel.[1,2] In a series of experiments stretching over approximately ten years, these women had discovered that the silver distribution in PbTe became uniform when the material was solidified in a centrifuge at a particular rotation rate. Such results would be expected only if buoyancy-driven convection were negligible compared to the freezing velocity at this rotation rate. On the other hand, since centrifugation increases the acceleration that appears in all of the equations of hydrodynamics, one would expect the convection to increase monotonically with increasing acceleration. Consequently, the results of Rodot and Regel were greeted with considerable skepticism by the crystal growth and hydrodynamic communities. In the Paris meetings we listed the possible influences of centrifugation on directional solidification, and discussed their potential relevance to the Rodot-Regel experiments. We decided that it would be useful to widen this discussion to all of those in the world who had interest and experience in materials processing during centrifugation.

The First International Workshop on Materials Processing at High Gravity was held in May of 1991 in Dubna, USSR, on the banks of the Volga River. The proceedings of this workshop was published in 1992 as a special issue of the Journal of Crystal Growth.[3]

In July of 1991, Regel came to Clarkson University for a 4-month visit as part of her collaborative research with Wilcox on solidification of InSb in the centrifuge. Subsequently she decided to remain at Clarkson, where she established the International Center for Gravity Materials Science and Applications. She also carried the primary responsibility for organizing the Second International Workshop on Materials Processing at High Gravity in Potsdam, NY, on the banks of the Racquette River. Like Dubna, the site of the first workshop, Potsdam is a village in a scenic area, concentrating on technology, with many foreign visitors, and a sizable river running through it. During this workshop, we dedicated Clarkson's unique centrifuge facility for materials processing research and related flow visualization.

Here we first discuss the phenomena that occur in centrifuges. Then we cover the state of our current understanding on directional solidification. Workshop papers on other techniques are described next, followed by hardware developments and recommendations for future research. We apologize for omission of relevant papers. Although our references are extensive here, we did not intended to include all prior literature.

PHENOMENA

When a fluid is placed into a centrifuge rotating at angular velocity ω, several changes occur:

1. The net acceleration is increased by the centrifugal acceleration $\omega^2 r$, where r is the distance from the axis of rotation.

2. The Coriolis acceleration $2\omega \times V$ is introduced, where $\times$ represents the vector cross product and V is the local fluid velocity in the rotating frame.

3. The acceleration vector varies in magnitude and direction throughout the fluid (because r in $\omega^2 r$ varies). We call this variation the "acceleration gradient."

In a centrifuge on earth, the net acceleration is the vector sum of earth's acceleration g and the centrifugal acceleration $\omega^2 r$, as shown in Figure 1.

The influence of increased net acceleration is usually as expected, while the Coriolis acceleration and the acceleration gradient often produce unexpected results. Increasing the net acceleration usually causes buoyancy-driven convection to increase, modifying heat and mass transfer throughout the system. For example, an increase in mass transfer increases the rate of growth of crystals from solutions and from the vapor.

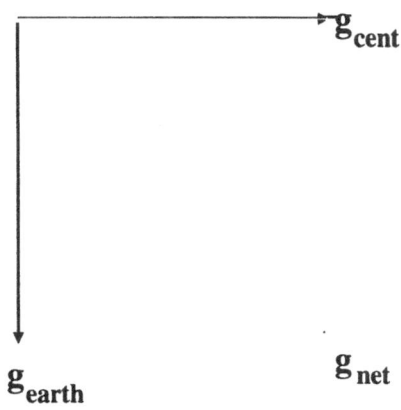

gcent

gearth   **g** net

Figure 1. Vector sum of earth's acceleration and centrifugal acceleration. Note that the magnitude and the direction of the net acceleration depend on both on the rotation rate ω and the radial position r.

Increased heat transfer in the melt during directional solidification by the gradient freeze technique can lower the temperature gradient in the melt, increase the freezing rate, and lead to morphological breakdown of the solid-liquid interface. The position and shape of the solid-liquid interface can be changed, thereby influencing impurity incorporation, thermal stress in the solid, and propagation of dislocations and grain boundaries.

Increasing the net acceleration also increases the sedimentation of second phase material, such as particles and bubbles. To the extent that particles cause nucleation of new grains, enhanced sedimentation of foreign particles can alter the grain size of solidified materials and decrease the spurious nucleation of new crystals during vapor and solution growth. Rise of bubbles during solidification would reduce the incorporation of gas bubbles throughout the solid and at the ampoule walls.

At very high accelerations, sedimentation of dissolved constituents occurs. This can be used, for example, to cause crystallization.

The "weight" of a material is increased by increasing the net acceleration. For a liquid, this can cause more intimate contact with the surface of its container, especially if the surface is rough. A solid may be plastically deformed by the force arising from its own weight.

The Coriolis acceleration modifies the flow pattern in fluids and the stability of the flow, but has little influence on the vigor of buoyancy-driven convection. Only rarely does the acceleration gradient have a strong influence on buoyancy-driven convection, but sometimes it assumes great importance. We will discuss this in some detail later.

DIRECTIONAL SOLIDIFICATION

The term "directional solidification" indicates that freezing occurs primarily in one direction. Many different techniques can be employed to accomplish this. In centrifuges, the techniques that have been used are zone melting and the gradient freeze technique. As described below, several different orientations between the total acceleration vector and the direction of freezing have been employed. We have learned that this orientation has a strong influence on the nature of the convection in the melt and on solidification.

The acceleration used for directional solidification experiments has usually been below 10 g. In this section we consider only experiments performed below 21 g.

Inverted gradient freeze technique

Müller and coworkers performed an extensive set of solidification experiments on InSb, temperature measurements in liquid metals, and theoretical modeling for the inverted gradient freeze technique shown in Figure 2.[4-11] The ampoule containing the material was placed in a furnace with the temperature decreasing with height. Solidification was made to proceed downward by slowly decreasing the furnace temperature. The furnace was attached to the centrifuge with a fixed angle, so that the resultant acceleration vector was aligned with the ampoule axis at only one rotation rate.

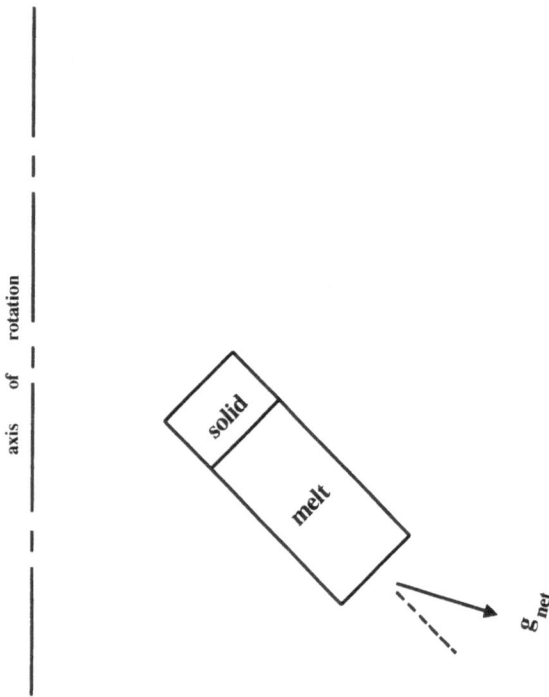

Figure 2. Inverted gradient freeze technique, with resultant acceleration misaligned with the ampoule axis.[4-11]

As the rotation rate was increased, the axial temperature gradient in the melt declined, indicating increased buoyancy-driven convection. Temperature fluctuations began to occur in the melt. Etching revealed impurity striations in the resulting crystals, indicating a fluctuating freezing rate. As the rotation rate was increased farther, the temperature fluctuations increased in magnitude until a critical rate was reached at which the temperature again became steady and the crystals no longer had striations. Müller and coworkers showed that this transition corresponds to a change in flow direction due to the Coriolis acceleration. Below the critical rotation rate, a single flow cell circulates in one direction and is irregular. Above the critical rotation rate, a single flow cell circulates in the opposite direction. Temperature fluctuations did not recur when the rotation rate was decreased, i.e. there was a hysteresis in flow behavior.

Chevy et al.[12] performed similar experiments, except that the furnace was attached to the centrifuge by a hinge, so that the net acceleration was always aligned with the ampoule axis, as shown in Figure 3. In this case, there was no hysteresis in flow behavior. As the rotation rate was increased, temperature fluctuations stopped at a critical rate, and resumed below this value when rotation was decreased. The acceleration required to stop the temperature fluctuations increased rapidly as the ampoule diameter was increased. The axial temperature gradient in the melt decreased dramatically as acceleration was increased, and was much less than the gradient in the furnace. This decrease in axial temperature gradient indicates vigorous convection.

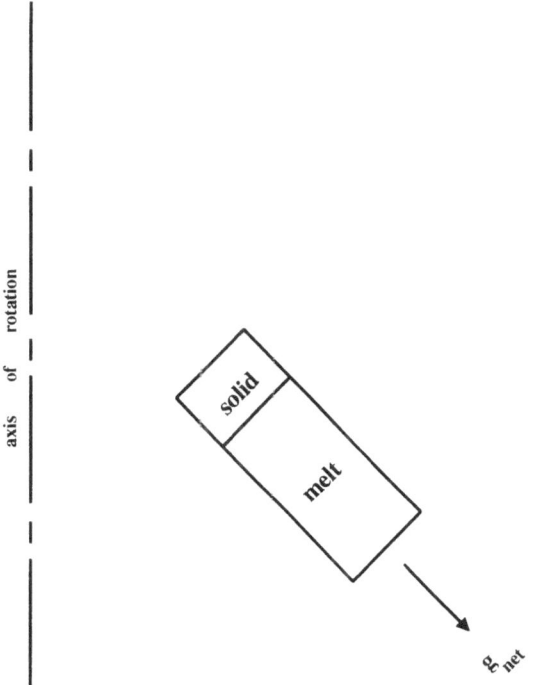

Figure 3. Inverted gradient freeze technique, with resultant acceleration aligned with the ampoule axis.

Horizontal gradient freeze technique

Gallium arsenide was solidified by the gradient freeze technique in a horizontal boat, attached to a centrifuge by a hinge so that the resultant acceleration was always normal to the ampoule axis and the surface of the melt.[13] This geometry is shown in Figure 4. Impurity striations were found in the resulting crystals, indicating fluctuations in freezing rate due to fluctuations in the convection. These striations diminished as the acceleration was increased.

Temperature measurements were made in liquid tin in the horizontal boat configuration with a temperature gradient down the furnace.[13,14] As the acceleration was increased, the axial temperature gradient in the molten tin decreased, indicating increased convection. Temperature fluctuations were observed in the melt. These temperature fluctuations decreased considerably with increasing acceleration if the centrifuge rotation was in the same sense as the convection roll in the melt. If the centrifuge rotation was in the opposite direction, then the temperature fluctuations increased with increasing rotation rate. A numerical model[15] of buoyancy-driven convection in the melt agreed with the experimental results. If the Coriolis force pushes the top and bottom streamlines closer together, then the flow is destabilized. If the Coriolis force pushes the streamlines apart, then the flow is stabilized and the temperature fluctuations should decrease with increasing rotation rate.

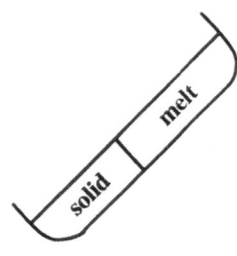

Top view

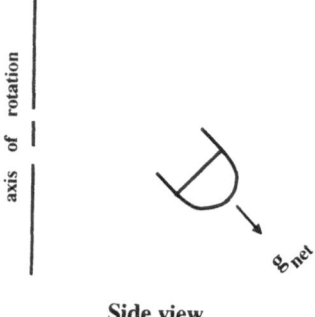

Side view

Figure 4. Directional solidification using a horizontal boat.

Gradient freeze technique

Regel and Rodot pioneered centrifugal directional solidification in the normal gradient freeze orientation shown in Figure 5.[1,2] In the horizontal boat method and the inverted gradient freeze technique, significant buoyancy-driven convection is expected. In the normal gradient freeze orientation at 1 g there would be no convection at all if there were no horizontal temperature gradients. However some radial temperature gradients are inevitable, and so gentle convection does occur.[16-21]

As mentioned earlier, Regel and Rodot directionally solidified Ag-doped PbTe in a gradient freeze furnace.[1,2] The ampoule axis was aligned with the net acceleration vector. The Ag concentration was measured along the centerline of each ingot, but not near the ends. Ingots solidified in the 18 m arm centrifuge at the Gagarin Cosmonaut Training Center outside Moscow had an uniform Ag concentration when the net acceleration was 5.2 g. As the acceleration deviated more and more from 5.2 g, the Ag concentration became less and less constant.

Similar results were obtained for PbTe in the 5.5 m arm centrifuge at Nantes in France, except that an uniform Ag concentration was obtained at about 2 g. The rotation rate for an uniform concentration was approximately the same in the Russian and French centrifuges.

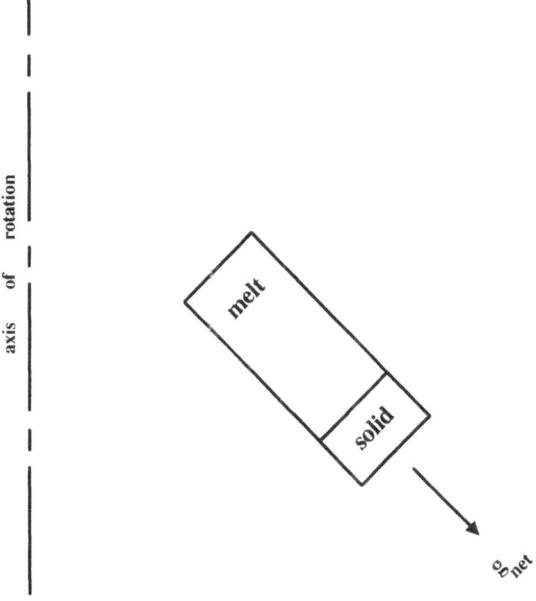

Figure 5. Vertical gradient freeze crystal growth in the centrifuge.

Theoretically, an uniform doping concentration is expected in the center portion of an ingot only when there is no convection in the melt during solidification. Other materials solidified in the Soviet centrifuge also exhibited behavior expected for reduced convection as the rotation rate increased.[22-31] These results were greeted by the scientific community with considerable skepticism. The existing theory for buoyancy-driven convection without rotation predicted that the vigor of the convection should increase monotonically with increasing acceleration. Recent theoretical work has provided considerable insight into convection in the vertical gradient freeze technique and an explanation for these experimental results.[32-39] Note particularly the papers by Arnold[36] and by Urpin[38] in this volume. The following is based on their work.

If density ρ decreases with height ($g \cdot \rho < 0$), buoyancy-driven convection occurs only when the acceleration vector is not perfectly aligned with the density gradient ($g \times \nabla \rho \neq 0$). When the acceleration vector is parallel to the ampoule axis, as in a vertical ampoule on earth, then convection occurs whenever a horizontal density gradient is present. In the absence of concentration gradients, the freezing interface is an isotherm, and the density gradient is perpendicular to the interface. Thus the driving force for convection near the freezing interface is directly related to the curvature of this interface. Indeed, from his numerical simulation of gradient freeze growth of germanium, Motakef found that the maximum velocity in the melt is proportional to the interface deflection and to the axial temperature gradient along the wall.[24] In the gradient freeze technique, the interface is concave,[16] and so convection is always expected on earth.

In a centrifuge, the acceleration vector is no longer parallel everywhere to the ampoule axis. Under some conditions, the acceleration can become normal to the concave interface in a gradient freeze experiment. When this happens, $g \times \nabla \rho = 0$, and there is no driving force for convection in the neighborhood of the interface.[35,36,38] We believe this is what happened in the gradient freeze centrifuge experiments described above. As the rotation rate was increased, the acceleration vector became more and more perpendicular to the concave freezing interface. For Ag-doped PbTe, we believe it became very nearly perpendicular at one particular rotation rate, as shown in Figure 6, and then deviated from this condition as the rotation was increased farther.

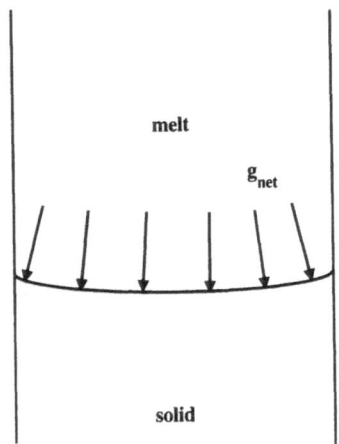

Figure 6.   Net acceleration vector normal to freezing interface in centrifuge.

We believe it is highly unlikely that convection was completely absent when an uniform Ag concentration was obtained in PbTe. The acceleration could not have been precisely aligned with the density gradient everywhere in the melt. So some convection would always have been present. Nevertheless, at the moderately high freezing rate used for the PbTe experiments, an uniform axial profile could have been obtained nonetheless. When the ratio of freezing rate to diffusion coefficient is sufficiently high, we can obtaine a concentration profile corresponding to the absence of convection even when some convection is present. Such conditions can yield an uniform axial concentration profile, while at the same time producing substantial radial variations in doping. In the PbTe experiments, the Ag concentration was measured only along the centerline, and so cross sectional variations would not have been observed. Similarly, the interface shape and the freezing rate were not measured, making comparison with theory impossible.

As noted above, the influence of convection on impurity segregation decreases as the freezing rate V increases.[e.g.21] The parameter of importance is the ratio V/D, where D is the diffusion coefficient of the impurity in the melt. If V/D is large compared to the convection, an axial concentration profile can be obtained that is characteristic of that expected theoretically in the absence of convection. In other words, obtaining such an axial concentration profile does not prove the absence of convection. A more sensitive measure of convection is the cross sectional variation in dopant concentration, which passes through a maximum with increasing freezing rate.[e.g.21]

No impurity striations could be found in InSb[40-43], GaSb[27] or Te[45] ingots prepared by the gradient freeze technique using either the 18 m Star City centrifuge or the 5.5 m Nantes centrifuge. These negative results indicate that the freezing rate did not fluctuate and that the convection was steady.

8

Table 1. Current understanding of solidification of semiconductors or metals by the gradient freeze technique in a centrifuge, as gained from experiments and theoretical modelling of convection in the melt.

Behavior	Vertical (freezing up; Fig. 5)	Inverted (freezing down; Fig. 3)	Horizontal (Fig. 4)
Convection	•Weak •Minimum versus g	•Strong •Increases with g	•Strong •Increases with g
Temperature	•Steady	•Fluctuations that cease above critical g	•Fluctuates. Influence of g depends on direction of rotation of centrifuge.
Temperature gradient in ampoule	•Negligible influence of g if furnace temperature profile is fixed	•Decreases with g	•Decreases with g
Freezing rate	•No influence of g if furnace temperature profile is fixed	•Increases with g if same furnace cooling rate and profile used	•Increases with g if same furnace cooling rate and profile used
Striations in crystal	•None. No influence of g	•None above critical g for cessation of temperature fluctuations	•Increases or decreases with g depending on direction of rotation of centrifuge
Bubbles and voids in crystal	•Fewer as acceleration increased		
Crystal quality	•Influenced by g, usually for unknown reasons.		

Temperature measurements in molten Sn, 70%Sb-30%Bi, Ge and Al revealed no fluctuations and no change in axial temperature gradient as the acceleration was increased.[12]

The primary dendrite arm spacing in Pb-Sn alloys decreased significantly as acceleration was increased.[46] Such behavior corresponds to reduced convection.

The variation in charge carrier concentration down Te crystals corresponded to the presence of some convection.[45] This variation reached a maximum as the acceleration was increased and then decreased at higher g. This behavior was attributed to the peculiar properties of molten Te, which shows a minimum in density as temperature is increased beyond the melting point.

$Pb_{0.8}Sn_{0.2}Te$ was directionally solidified in several configurations.[47,48] Solidification upward at 1 g produced compositional profiles corresponding to those expected in the presence of moderately strong convection. Solidification downward by the inverted Bridgman method indicated that convection was reduced. This was explained by the dominance of solutal effects, i.e. the density of the melt was influenced more by composition gradients than by temperature gradients. Gradient freeze growth in centrifuges yielded complex composition profiles, in both the inverted and normal orientations. This behavior may indicate a strong variation in freezing rate and/or convection during solidification in the centrifuge.

Influence of centrifugation on microstructure and perfection

We have concentrated thus far on the influence of centrifugation on compositional variations in the resulting materials due to changes in convection in the melt during solidification. We believe we have gained a fairly good understanding of the influence of centrifugation on buoyancy-driven convection, as summarized in Table 1. Centrifugation has also been found to influence the microstructure and perfection of directionally solidified materials. In most cases, we do not know why.

There are at least three possible explanations for an improvement of microstructure. First, foreign particles may sediment out during centrifugation prior to solidification, so that they cannot nucleate new grains and twins during solidification. Second, by altering the heat transfer in the system, centrifugation may cause the interface to become more favorable for grain selection, i.e. less concave. Third, centrifugation causes gas bubbles in the melt to float away from the freezing interface. (Many fewer gas bubbles are present on the surface and in the interior of materials solidified in the centrifuge.) It is possible that gas bubbles at the freezing interface cause nucleation of new grains or twins. For example, a bubble may suddenly move, altering the heat transfer in the neighborhood of the interface and causing very rapid solidification at that location.

Following are some examples of the influence of centrifugation at moderate accelerations on microstructure and perfection. In their gradient freeze experiments on Ag-doped PbTe, Rodot and Regel found that only the ingot solidified at 5.2 g was a high quality single crystal.[1,2] Experiments on gradient freeze solidification of germanium in the centrifuge also showed improved grain size as the rotation rate was increased.[49,50] The solid-liquid interface of GaSb solidified in the Star City centrifuge was flatter than when GaSb was solidified in space or at 1 g.[27] The number of gas bubbles in InSb decreased as the acceleration increased.[40-43] Although there were differences in the mobility and numbers of grain and twin boundaries between the InSb ingots, no trend with acceleration could be discerned.[40-44] The mobility of holes in Te was relatively constant when it was prepared by the gradient freeze technique at 1 g.[45] Te solidified at 5 and 10 g had a significantly lower mobility that increased as one moved down the ingots. This behavior was attributed to increased convection in the melt during solidification that increased the disorder in the resulting crystals.

As noted earlier, in the solidification of PbTe-SnTe alloys, convection is greater in the normal orientation than in the inverted orientation.[48] Increasing acceleration in the normal gradient freeze orientation caused formation of a cellular structure, metallic inclusions, and nucleation of new grains.[47] This was attributed to increased convection, leading to a reduced temperature gradient and an increased freezing rate, both of which increase the likelihood of morphological instability.[47]

In the present volume, it is speculated that increased acceleration might increase the sticking of crystals to the ampoule wall, thereby increasing plastic deformation due to the difference in thermal expansion coefficient between the crystal and the ampoule.[51] Acceleration was observed to increase the dislocation density of GaAs grown by the horizontal gradient freeze technique in sand-blasted silica boats.[13] It was speculated that the increased weight forced the melt into the pits and valleys of the boat wall, thereby increasing the adhesion of the resulting solid. One might also expect increased weight of the melt and the solid to force the solid to deform plastically in order to remain in contact as it cools from the melting point. (Normally the ampoule is coated with a non-stick coating to assist the solid in breaking away from the wall during cooling.)

OTHER RESEARCH ON MATERIALS PROCESSING IN CENTRIFUGES

Many other materials processing operations have been carried out at moderate accelerations. For example, a high TC superconductor mixture of Bi-Sr-Ca-Cu-O was crystallized in the vertical gradient freeze orientation.[52] The superconducting transition temperature increased as one moved down the crystallized mixture, reaching as high as 130 K. Apparently crystals sedimented as crystallization proceeded.

GeSe crystals were grown by the vapor transport technique in the normal and inverted gradient freeze orientations using the Star City centrifuge.[53] At 1 g, the net transport rate was only slightly higher for the inverted orientation, which would be expected to produce more vigorous convection in the gas. However there were fewer but larger crystals from the normal orientation. Centrifugation at 10 g in the normal orientation increased the transport rate by about 35%. In the inverted orientation, the transport rate was roughly proportional to g. The deposition pattern in both orientations was nearly axisymmetric at 1 g and become very asymmetric at 10 g. The largest crystals, with well defined Laue patterns, were obtained at 10 g in the inverted orientation.

Several papers on other topics appear in the present volume. GaAs was chemically vapor deposited onto a GaAs hemispherical substrate.[54] The growth rate was proportional to $g^{1/4}$ from 1 to 10 g, with more spurious nucleation on the ampoule wall as g was increased. A numerical model was developed to explain the results. The flame height of a fire[55] varied as $g^{-1/3}$, the fluctuation frequency as $g^{1/2}$, and the total radiant power as $g^{-0.3}$. The experimental results were correlated using the predictions of numerical models and scaling analysis. In a numerical study of welding, increased acceleration was predicted to enhance the convection in the weld pool, thereby influencing the heat transfer, the depth and width of the two phase region, and the pool depth-to-width ratio.[56] An experimental study is reported on the influence of centrifugation, sample orientation and heater power on transient heat transfer from a wire -- a technique proposed for the measurement of the thermal diffusivity of semiconductor melts.[57] Two papers deal with a system in which the axis of rotation is normal to a semi-infinite growth surface.[58,59] One paper considers cellular convection due to a concentration gradient such that the density increases with height.[58] In the other, the morphological stability of a freezing interface was considered with density depending on both concentration and temperature.[59]

Finally, we consider processing in ultracentrifuges, in which the acceleration can reach several hundred thousand times earth's gravity g. Sedimentation of dissolved constituents becomes appreciable.[e.g. 58,59] This phenomenon has been used to move solvent inclusions through crystals[60-62] and to cause crystals to grow from solution.[63-65] In the present volume it is reported that centrifugation caused TNT and RDX crystals to grow free of voids.[65] Separation of the constituents of eutectic metal mixtures was demonstrated at very high accelerations.[66-68]

Sedimentation of the gel phase has proved helpful in the precipitation of zeolites.[69-71] In the present volume is described experiments and theory for gel polymerization by light in a centrifuge.[72] As g was increased, the radial variation in properties of the polymer increased, including Young's modulus and pore size. The effect depended on when centrifugation was begun during the polymerization. Very high acceleration caused mechanical destruction of the polymer.

FACILITIES

Moderate g experiments were first performed by Müller et al. in a centrifuge facility constructed at Erlangen in Germany.[4-11] The furnace was attached at a fixed angle to the centrifuge axis. The early Rodot-Regel experiments were performed on the 18 m radius centrifuge at Star City, USSR.[1,2,26] Their gradient freeze furnace was constructed at CNRS Meudon in France. Because the primary purpose of the centrifuge was training of cosmonauts, it was available for only a few days each year for materials processing experiments. Similarly, only limited time was available for subsequent experiments on the centrifuge at Nantes, France.[73] In the present volume are described two centrifuge facilities fabricated in the United States specifically for materials processing research,[45,74] as well as a large, general purpose centrifuge in Canada.[75] Currently the Clarkson centrifuge[74] is being used for gradient freeze experiments on InSb and CdTe. It was also designed to accommodate flow visualization experiments. It is worth noting that the cost is modest for construction of large materials processing centrifuges useful up to 20 g.

FUTURE RESEARCH

There are two motives for studying materials processing in centrifuges. First, such research improves our understanding of the influence of acceleration and convection on materials processing. Second, there are commercial opportunities for production of unique and improved materials that cannot be prepared under normal earth conditions or in space.

Additional careful experiments are needed in gradient freeze solidification, with materials of various types, including semiconductors, metal alloys, organic compounds and oxides. A wide range of freezing rates, temperature gradients, and compositions should be investigated. When possible, seeding should be used to provide single crystals of desired orientations. Interface demarcation should be used so that the interface shape and freezing rate versus position are known. The resulting ingots should be characterized more thoroughly than in the past, including, for example, analysis of impurity concentration to the ends of the crystals and over entire cross sections, microstructure, dislocation content, inclusions, and electrical properties.

Similarly, additional research should be performed on solution crystal growth, vapor transport, chemical vapor deposition, polymerization, and welding. Other opportunities exist in Bridgman-Stockbarger solidification, electrodeposition, fabrication and joining of composite materials, and fine particle processing.

Flow visualization and temperature measurements should be performed and compared with theoretical predictions.

We plan to carry out much of the above research here at Clarkson University.

Acknowledgments

The Second International Workshop on Materials Processing at High Gravity was supported by a grant from the United States National Science Foundation. We are grateful to Clarkson University for arranging the dedication of our centrifuge facility during the workshop, as well as television interviews of participants. Ramnath Derebail prepared the figures used in this introduction.

REFERENCES

1. H. Rodot, L.L. Regel, G.V. Sarafanov, H. Hamidi, I.V. Videskii, and A.M. Turtchaninov, Cristaux de tellurure de plomb elabores en centrifugeuse, *J. Crystal Growth* 79:77 (1986).
2. H. Rodot, L.L. Regel, and A.M. Turtchaninov, Crystal growth of IV-VI semiconductors in a centrifuge, *J. Crystal Growth* 104:280 (1990).
3. L.L. Regel, M. Rodot, and W.R. Wilcox, editors, "Material Processing in High Gravity, Proceedings of the First International Workshop on Material Processing in High Gravity," North-Holland, Amsterdam (1992). Also volume 119 of the Journal of Crystal Growth.
4. G. Müller, E. Schmidt, and P. Kyr, Investigation of convection in melts and crystal growth under large inertial accelerations, *J. Crystal Growth* 49:387 (1980).
5. G. Müller, Crystal growth at greater than 1 g, *in*: "ESA Special Publication No. 114," European Space Agency, Paris (1980) pp 213-216.
6. G. Müller and G. Neumann, Suppression of doping striations in zone melting of InSb by enhanced convection on a centrifuge, *J. Crystal Growth* 59:548 (1982).
7. G. Müller, Convection in melts and crystal growth, *in*: "Convective Transport and Instability Phenomena," J. Zierep and H. Oertel, Jr., eds., Braun Verlag, Karlsruhe (1982).
8. G. Müller, "Über die Enstehung von Inhomogenitaten in Halbleiterkristallen bei der Herstellung aus Schmelzen," Selisch Fachbuch-Verlag, Langensendelbach (1986) pp 151-165.
9. G. Müller, A comparative study of crystal growth phenomena under reduced and enhanced gravity, *J. Crystal Growth* 99:1242 (1990).
10. W. Weber, G. Neumann, and G. Müller, Stabilizing influence of the Coriolis force during melt growth on a centrifuge, *J. Crystal Growth* 100:145 (1990).
11. G. Müller, G. Neumann, and W. Weber, The growth of homogeneous semiconductor crystals in a centrifuge by the stabilizing influence of the Coriolis force, *J. Crystal Growth* 119:8 (1992).
12. A. Chevy, P. Williams, M. Rodot, and G. Labrosse, Removal of convective instabilities in liquid metals by centrifugation, present volume.
13. B. Zhou, F. Cao, L. Lin, W. Ma, Y. Zheng, F. Tao, and M. Xue, Growth of GaAs at high gravity, present volume.
14. W.J. Ma, F. Tao, Y. Zheng, M.L. Xue, B.J.Zhou, and L.Y. Lin, Response of temperature oscillations in a tin melt to centrifugal effects, present volume.
15. F. Tao, Y. Zheng, W.J. Ma, and M.L. Xue, Unsteady thermal convection of melts in a 2-D horizontal boat in a centrifugal field with consideration of the Coriolis effect, present volume.
16. C.E. Chang, V.F.S. Yip, and W.R. Wilcox, Vertical gradient freeze growth of gallium arsenide and naphthalene: theory and practice, *J. Crystal Growth* 22:247 (1974).
17. C.E. Chang and W.R. Wilcox, Control of interface shape in the vertical Bridgman-Stockbarger technique, *J. Crystal Growth* 21:135 (1974).
18. S. Sen and W.R. Wilcox, Influence of crucible on interface shape, position and sensitivity in the vertical Bridgman-Stockbarger technique, *J. Crystal Growth* 28:36 (1975).
19. T.W. Fu and W.R. Wilcox, Influence of insulation on stability of interface shape and position in the vertical Bridgman-Stockbarger technique, *J. Crystal Growth* 48:416 (1980).
20. G.T. Neugebauer and W.R. Wilcox, Convection in the vertical Bridgman-Stockbarger technique, *J. Crystal Growth* 89:143 (1988).
21. S. Motakef, Interference of buoyancy-induced convection with segregation during directional solidification: scaling laws, *J. Crystal Growth* 102:197 (1990).
22. L.L. Regel et al., Effect of increased gravity on the structure of directionally solidified aluminum-copper eutectic, *Fiz. Khim. Obrab. Mater.* 45 (1989).
23. L.L. Regel, A.M. Turchaninov, R.V. Parfeniev, I. Farbshtein, N.K. Shulga, S.V. Nikitin, and S.V.

Yakimov, "Electrofizicheskie Svoictva Monokristallov Tellura i Splava Te$_{1-x}$Se$_x$, Poluchennikh v Usloviyakh pri Vishennoi Gravitatsii (5 g$_o$ i 10 g$_o$)," USSR Space Research Institute, Moscow (July 1989).

24. L.L. Regel, I.V. Videnskii, V.V. Zubenko, I.M. Cafonova, and I.V. Telegina, Vliyanie povishennoi gravitatsii na strukturu napravlenno - za kristallizovannik evtektik alyominii - medi, *Fizika i Chimiya Obrabotki Materialov* 23:45 (1989).

25. P. Bartsi, L.L. Regel, and I. Solyom, *in*: "Proceedings of the 4th Intercosmos Seminar on Cosmic Materials and Technologies," Bucharest (1989) pp 117-137.

26. B.V. Burdin, L.L. Regel, A.M. Turchaninov, and O.V. Shumaev, The peculiarities of material crystallization experiments on the CF-18 centrifuge in high gravity, *J. Crystal Growth* 119:61 (1992).

27. L.L. Regel and O.V. Shumaev, GaSb directional solidification in high gravity conditions, *J. Crystal Growth* 119:70 (1992).

28. P. Barczy, J. Solyom, and L.L. Regel, Solidification AlNi(Cu) eutectics at high gravity, *J. Crystal Growth* 119:160 (1992).

29. L.L. Regel et al., Te and Te-Se alloy crystal growth under higher gravity, *J. Phys. France* 2:373 (1992).

30. Z. Chvoj and C. Barta, Remark on the influence of gravitation on the solidification of the binary systems, *Czech. J. Phys. B* 36:868 (1986).

31. C. Barta, F. Fendrych, E. Krcova, and A. Triska, Directional solidification of complex-forming eutectic melt of the lead dichloride - silver chloride dielectric system under conditions of zero, normal and increased gravity, *Adv. Space Res.* 8:167 (1988). Also, *in*: "Proceedings of the 4th Intercosmos Seminar on Cosmic Materials and Technologies," V. Lupei and D.Toma, eds., Rumanian Academy of Science, Bucharest (1989).

32. W. Arnold, W.R. Wilcox, F. Carlson, A. Chait, and L.L. Regel, Transport modes during crystal growth in centrifuge, *J. Crystal Growth* 119:24 (1992).

33. W.A. Arnold, W.R. Wilcox, F. Carlson, L.L. Regel, and A. Chait, Flow mode transitions during crystal growth in a centrifuge, *J. Crystal Growth* (submitted).

34. W. Arnold, W. Wilcox, F. Carlson, A. Chait, and L. Regel, Crystal growth of semiconductor compounds in a centrifuge, *in*: "Proceedings of the Society of Engineering Science," Gainesville (November 1991).

35. W. Arnold, "Numerical Modeling of Directional Solidification in a Centrifuge," PhD Thesis, Clarkson University (1993).

36. W.A. Arnold and L.L. Regel, Thermal stability and the suppression of convection in a rotating fluid on earth, present volume.

37. M.A. Fikri, G. Labrosse, and M. Betrouni, The melt phase hydrodynamics for the "stabilized" Bridgman procedure applied under centrifugation; preliminary analysis and numerical results, *J. Crystal Growth* 119, 41-60 (1992).

38. V.A. Urpin, Convective flows during crystal growth in a centrifuge, present volume.

39. P.A. Vorobiov, N.A. Baturin, and O.V. Shumaev, Laminar convection in the melt during crystal growth in a centrifuge, *J. Crystal Growth* 119:111 (1992).

40. R. Derebail, W.R. Wilcox, and L.L.Regel, Directional solidification of InSb in a centrifuge, *J. Crystal Growth* 119:98 (1992).

41. R. Derebail, W.R. Wilcox, and L.L. Regel, The influence of gravity on the directional solidification of indium antimonide, *J. Spacecraft & Rockets* 30:202 (1993).

42. R. Derebail, "Study of Directional Solidification of InSb under Low, Normal and High Gravity," M.S. Thesis, Clarkson University (1990).

43. R. Derebail, "Directional Solidification of InSb in the Centrifuge," PhD Thesis, Clarkson University (1994).

44. L.I. Farbshtein, R.V. Parfeniev, S.V. Yakimov, L.L. Regel, R. Derebail, and W.R. Wilcox, Analysis of impurity distribution by galvanomagnetic method in InSb obtained under high gravity conditions, present volume.

45. L.I. Farbshtein, R.V. Parfeniev, N.K. Shulga, and L.L. Regel, Variation of effective impurity segregation coefficient in tellurium grown under high gravity, present volume.

46. R.N. Grugel, A.B. Hmelo, C.C. Battaile, and T.G. Wang, Microstructural development in Pb-Sn alloys subjected to high gravity during controlled directional solidification, present volume.

47. L.L. Regel, A.M. Turchaninov, O.V. Shumaev, I.N. Bandeira, C.Y. An, and P.H.O. Rappl, Growth of lead-tin telluride crystals in high gravity, *J. Crystal Growth* 119:94 (1992).

48. Y.A. Chen, I.N. Bandeira, A.H. Franzan, S. Eleutério Filho, and M.R. Slomka, The influence of gravity on Pb$_{1-x}$Sn$_x$Te crystals grown by the vertical Bridgman method, present volume.

49. A. Chevy, Cristallogenese du germanium en centrifugeuse, *Compte Rendue Acad. Sci. Paris* 307:1147 (1988).

50. A. Chevy, Private Communication, Universite Pierre et Marie Curie, Paris, France (1990).

51. T. Lee, J.C. Moosbrugger, F.M. Carlson, and D.J. Larson, Jr., The role of thermal stress in vertical Bridgman growth of CdZnTe crystals, present volume.

52. M.P. Volkov, B.T. Melekh, R.V. Parfeniev, N.F. Kartenko, and L.L. Regel, Properties of superconducting Bi-Sr-Ca-Cu-O system remelted under high gravity conditions, *J. Crystal Growth* 119:122 (1992).

53. H. Wiedemeier, L.L. Regel, and W. Palosz, Vapor transport and crystal growth of GeSe under normal and high acceleration, *J. Crystal Growth* 119:79 (1992).

54. J.C. Launay, S. Bouchet, A. Randriamampianina, P. Bontoux, and P. Gibart, Epitaxial growth on a GaAs hemisphere substrate at 1 g and under hypergravity, present volume.

55. J. Chen, J.M. Most, P. Joulain, and D. Durox, Fire behavior in macrogravity, present volume.

56. J. Domey, D.K. Aidun, G. Ahmadi, L.L. Regel, and W.R. Wilcox, Numerical simulation of the effect of gravity on weld pool shape, present volume.

57. T. Hibiya, S. Nakamura, K.W. Yi, and K. Kakimoto, Coriolis effect on heat transfer experiment using hot-wire technique on centrifuge, present volume.

58. K.O. Pedersen, Uber das Sedimentationsgleichgewicht von anorganischen Salzen in der Ultrazentrifuge, *Z. Phys. Chem.* A170:41 (1934).

59. D.J. Cox, Computer simulation of sedimentation in the ultracentrifuge. III. Concentration-dependent sedimentation, *Arch. Biochem. Biophys.* 119:230 (1967).

60. W.R. Wilcox and P. Shlichta, Movement of crystal inclusions in a centrifugal field, *J. Appl. Phys.* 42:1823 (1971).

61. W.R. Wilcox, Movement of liquid inclusions by centrifugation, *J. Crystal Growth* 13/14:787 (1972).

62. T.R. Anthony and H.E. Cline, The kinetics of droplet migration in solids in an accelerational field, *Phil. Mag.* 22:893 (1970).

63. P.J. Shlichta, Crystal growth and materials processing above 1000 g, *J. Crystal Growth* 119:1 (1992).

64. P.J. Shlichta and R.E. Knox, Growth of crystals by centrifugation, *J. Crystal Growth* 3/4:808 (1968).

65. M.Y.D. Lanzerotti, J. Autera, J. Pinto, and J. Sharma, Crystal growth of energetic materials during high acceleration using an ultracentrifuge, present volume.

66. R.S. Sokolowski, "Gravitational influence on binary alloy melt equilibria and eutectic solidification," Ph.D. Thesis, Rensselaer Polytechnic Institute, Troy, NY (1981).

67. M.E. Glicksman and R.S. Sokolowski, Gravitational influence on binary alloy melt equilibria. *Adv. Space Res.* 3:129 (1983).

68. R.S. Sokolowski and M.E. Glicksman, Gravitational influence on eutectic solidification, *J. Crystal Growth* 119:126 (1992).

69. D.T. Hayhurst, P.J. Melling, W.J. Kim, and W. Bibbey, *in:* "Zeolite Synthesis," M.L. Occelli and H.E. Robson, eds., American Chemical Society (1989) ch 17.

70. W.J. Kim, "The Effect of Elevated Gravity on the Crystallization of the MFI Zeolites, ZSM-5 and Silicalite," Ph.D. Thesis, Cleveland State University, Cleveland, Ohio (1989); through *Chem. Abstr.* 112:219459 (1990).

71. D.T. Hayhurst, W.J. Kim, and P.J. Melling, "Crystal Growth in Enhanced Gravitational field." US Patent Application 233,287 (1988); PCT Int. Appl. WO 90 02,221 (1990); through *Chem. Abstr.* 113:32438 (1990).

72. V.A. Briskman, K.G. Kostarev, and T.P. Lyubimova, Gel polymerization at high gravity, present volume.

73. J. Garnier and L.M. Cottineau, Questions raised about material processing in a centrifuge: lessons derived from the LCPC's experience, *J. Crystal Growth* 119:66 (1992).

74. R. Derebail, W.A. Arnold, G.J. Rosen, W.R. Wilcox, and L.L. Regel, HIRB - the centrifuge facility at Clarkson, present volume.

75. M.J. Paulin, R. Phillips, J.I. Clark, R. Meaney, D. Millan, and K. Tuff, Establishment of the new C-CORE centrifuge center, present volume.

THERMAL STABILITY AND THE SUPPRESSION OF CONVECTION IN A ROTATING FLUID ON EARTH

William A. Arnold and Liya L. Regel

International Center for Gravity Materials Science and Applications
Clarkson University
Potsdam, NY 13699-5700

ABSTRACT

Thermal stability in a rotating fluid on earth is examined. Thermal stability refers here to the fluid state where convection is absent or at a minimum even in the presence of thermally induced density gradients. We examine the conditions which bring about thermal stability in a rotating fluid on earth through numerical simulations. It is shown that at least one thermal field exists for a rotating fluid with a gravitational background field where convection does not occur. The numerical model used is three-dimensional.

BACKGROUND

Many materials processing operations are influenced by convection. Bulk crystal growth of semiconductors from the melt is one such process. A common method for producing semiconductor crystals is directional solidification. In directional solidification, an ampule containing a charge is melted, then solidified from one end to the other. During this process, there is an interface between the melt and the solid. The interface is typically curved and isothermal. Horizontal temperature gradients cause buoyancy-driven convection.

Several directional solidification techniques are known to suppress convection. Solidification in the microgravity environment of space is one such method. Convection is reduced because the driving force of gravity is greatly reduced. The drawback here is that processing in space is very expensive. Another method is magnetic damping. When a strong magnetic field is present during solidification of conductors, the magnetic field retards convection.[1] The drawback here is that the level of convective damping depends on the electrical conductivity of the melt.

The fluid dynamics of rotating fluids can be complex. Rotating fluids can exhibit several unique and counter-intuitive qualities. Some examples have been known for a long time such as geostrophic flows[2] and Taylor columns.[2] Others are more recent, such as the stabilizing effect of the Coriolis force on buoyancy driven flow[3] and the recent experimental evidence[4] on the apparent suppression of convective transport at a well-defined acceleration level during solidification of semiconductor materials in a centrifuge. The last of the listed phenomena is of special interest as it is directly applicable to the crystal growth industry. The apparent suppression of convection leads to nearly uniform axial doping in the ingot. To date, two hypotheses have been proposed to explain the nearly uniform axial doping.[5] One of these hypotheses is that the convection is reduced to near zero at one particular rotation rate. Uniform axial doping would be expected only in the absence of convection.

The special acceleration level noted above depended on the arm length of the centrifuge used to directionally solidify Ag-doped PbTe ingots. It was shown that all ingots with uniform axial doping were processed at the same rotation rate. Thus, the controlling factor in convective suppression may be the rotation rate and not the acceleration level. Here we examine a convection suppression mechanism that may be working alone or in conjunction with other mechanisms during centrifugation.

Buoyancy-driven flow in a rotating fluid is much more complicated than in a non-rotating fluid (such as in a constant acceleration field, i.e. gravity) due to the spatially varying acceleration vector and the Coriolis acceleration. It is not surprising that several important and unique aspects of a rotating fluid where density gradients are present, have not yet been addressed. The most basic of these is the concept of thermal stability. By examining the thermal stability of a rotating fluid in an ideal configuration, the mechanism of convective suppression becomes evident. Although only thermally-induced density gradients are examined here, the mechanism extends to systems with solutally-induced density gradients.

THERMAL STABILITY THEORY IN NON-ROTATING AND ROTATING FLUIDS

The basic concept of a thermally stable fluid implies that there is no buoyancy-induced flow. We now undertake to specify the conditions under which this would occur in a rotating system. Our analysis begins by examining the conservation of momentum equation:

$$\rho \left(\frac{\partial}{\partial t} \vec{u} + \vec{u} \bullet \nabla \vec{u} \right) = -\nabla p + \rho \vec{g} + \mu \tilde{\nabla}^2 \vec{u} - \rho \left(\vec{\omega} \times \vec{\omega} \times \vec{R} \right) - \rho \left(2 \vec{\omega} \times \vec{u} \right) \qquad (1)$$

where ω is constant in time. In the absence of convection, equation 1 reduces to:

$$\nabla p = \rho \vec{g} - \rho \left(\vec{\omega} \times \vec{\omega} \times \vec{R} \right) \qquad (2)$$

Taking the curl of equation 2 with constant gravitational acceleration and rotation rate yields:

$$0 = \left(\nabla \rho \times \vec{g} \right) - \nabla \rho \times \left(\vec{\omega} \times \vec{\omega} \times \vec{R} \right) \qquad (3)$$

When equation 3 is applied to a fluid in a constant acceleration field, such as earth's gravitational field, thermal fields exist that are thermally stable. Such thermal fields have flat isotherms that are perpendicular to the gravitational field everywhere. That is, there is no horizontal temperature gradient. In addition, to apply at all Rayleigh numbers, the hotter fluid is above the cooler fluid with respect to the gravitational vector, assuming a positive coefficient of thermal expansion. A thermally stable temperature field for a constant acceleration in which no flow occurs is illustrated in figure 1. A more precise way of stating this is that the acceleration field is parallel to the density gradient at all points in the fluid, or that:

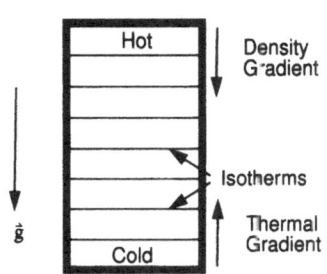

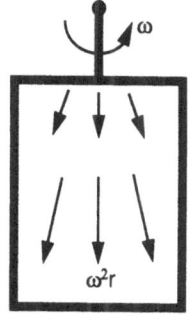

Figure 1. Thermally stable configuration for a constant acceleration field (denoted g).

Figure 2. Acceleration field for an enclosed rotating fluid without a constant background acceleration field.

$$\vec{g} \times \nabla \rho = 0 \quad \text{and} \quad [\vec{g} \bullet \nabla \rho] > 0 \tag{4}$$

This analysis assumes that the coefficient of thermal expansion is positive. Extending this concept and using the Boussinesq approximation, it is seen that the acceleration gradient is antiparallel to the thermal gradient at each and every point in the fluid, or in general that:

$$\vec{g} \times \nabla \rho_0 (1 - \beta \Delta T) = 0 \tag{5}$$

where $\Delta T = T - T_{ref}$. Even though β may be a function of temperature, equation 5, along with the relations:

$$\nabla \beta = (\frac{\partial \beta}{\partial T}) (\frac{\partial T}{\partial r} + \frac{\partial T}{\partial z} + \frac{1}{r}\frac{\partial T}{\partial \theta}) = (\frac{\partial \beta}{\partial T}) \nabla T \tag{6}$$

$$\nabla (\beta \Delta T) = \left[(\frac{\partial \beta}{\partial T}\Delta T) + \beta \right] \nabla T \tag{7}$$

and the assumption that $\beta > 0$ reduces to:

$$\vec{g} \times \nabla T = 0 \tag{8}$$

Likewise, the second part of equation 4 reduces to:

$$\vec{g} \bullet \nabla T < 0 \tag{9}$$

In a rotating fluid, the acceleration field is not homogeneous, i.e. the acceleration vector varies in magnitude and direction throughout the fluid. Application of the above analysis to equation 3 predicts that there still exists at least one family of thermal configurations in a rotating fluid where convective flow ceases. Here this state is called the thermally stable configuration. To illustrate this, figure 2 shows the centrifugal acceleration field in a centrifuge without the inclusion of a background gravitational acceleration. This scenario would be experienced by a rotating fluid in space (i.e., a centrifuge in space). The acceleration field always points radially out from the axis of rotation and the magnitude of the acceleration is proportional to the radial distance. In addition, without the complication of background acceleration, this field is two-dimensional. There is no acceleration along the axis of rotation. Thermal stability theory predicts that one family of thermal fields that result in an absence of convection have circular isotherms centered about the axis of rotation. The fluid is cooler as one moves radially outward from the axis of rotation, so that the hotter fluid is "over" the colder fluid in relation to the acceleration vector. Here the acceleration field is perpendicular to the isotherms everywhere.

With the inclusion of a constant background acceleration, the thermally stable field is not readily recognizable. This scenario occurs in centrifuges on earth where the background acceleration is earth's gravitational field. One family of thermal fields that lead to a thermally stable configuration have isotherms that are paraboloids centered on the axis of rotation, as shown in figure 3. In figure 3, a cylindrical coordinate system is shown. The explanation for the paraboloidal isotherms begins by examining the acceleration field, which is:

$$\vec{a} = \vec{g} + \omega^2 \hat{r} = -g\hat{z} + \omega^2 r\hat{r} \tag{10}$$

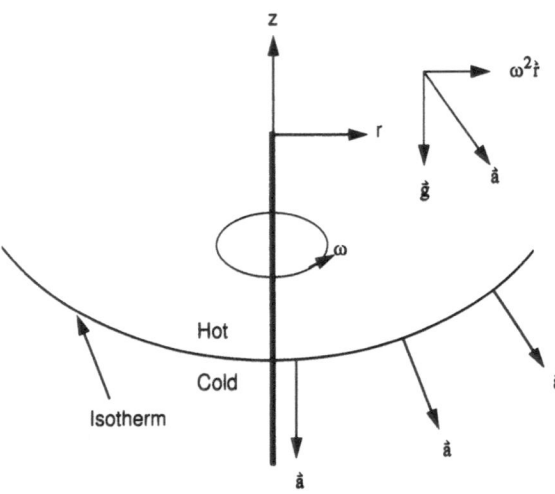

Figure 3. Thermally stable configuration for a rotating fluid with a constant background acceleration field (Cross-section through the axis of rotation shown).

The unit directional of the acceleration vector is:

$$\hat{e} = \frac{-g\hat{z} + \omega^2 r\hat{r}}{\sqrt{g^2 + \omega^4 r^2}} \tag{11}$$

A family of paraboloids centered about the z axis can be represented by the equation:

$$bz = d + cr^2 \tag{12}$$

where b, c and d are constants specific to each individual paraboloid. The outward unit normal to equation 12 is:

$$\hat{n} = \frac{-b\hat{z} + 2cr\hat{r}}{\sqrt{b^2 + 4c^2}} \tag{13}$$

which is in the same form as equation 11. Thus, the acceleration field in a centrifuge or earth is of a paraboloidal nature.

The intent of the present work is to numerically verify the thermal stability configuration for a rotating fluid on earth.

NUMERICAL VERIFICATION OF THE THERMALLY STABLE CONFIGURATION FOR A ROTATING FLUID ON EARTH

Governing Equations

The analysis here assumes an incompressible fluid with small thermally-induced density variations. The density is only allowed to vary in the body force term driving the flow. For constant gravitational acceleration and modest temperature gradients with a low coefficient of thermal expansion, the Boussinesq approximation is valid. When other body forces are incorporated (here through the centrifugal and Coriolis accelerations), the use of this approximation must be carefully decided upon. When density gradients are present the centrifugal force is nonconservative and must be solved for. This dependency is the result of the spatially non-constant centrifugal acceleration and the density gradients. However, when the product $\beta\Delta T_{max}$ is small, a modified Boussinesq approximation should be valid. The modified Boussinesq approximation accounts for density variations in the centrifugal term.

The governing conservation equations with the modified Boussinesq approximation applied to the convective terms and the Coriolis acceleration, in dimensional form with constant material properties, are:

$$\nabla \bullet \vec{u} = 0 \tag{14}$$

$$\rho_o (\frac{\partial}{\partial t}\vec{u} + \vec{u} \bullet \nabla \vec{u}) = -\nabla P - \rho_o \vec{g}\beta (T - T_{ref}) + \mu \tilde{\nabla}^2 \vec{u}$$
$$+ \rho_o \beta (T - T_{ref})(\vec{\omega} \times \vec{\omega} \times \vec{R}) - \rho_o (2\vec{\omega} \times \vec{u}) \tag{15}$$

$$\rho_o C_P (\frac{\partial}{\partial t}T + \vec{u} \bullet \nabla T) = k\nabla^2 T + q_s \tag{16}$$

where k and μ are non-temperature dependent, the viscous dissipation function has been neglected, $\vec{\omega}$ is constant in time, $\tilde{\nabla}^2$ is the vector Laplacian operator, and P is a term that includes the combined effect of local pressure and the static gravitational and centrifugal forces ($P = p + \rho_o gh - \rho_o \omega^2 r^2/2$). These equations are in a three-dimensional vector notation. The cylindrical coordinate system (r,θ,z) with rotation about the z axis was used for the two-dimensional numerical simulations and is used to interpret this analysis. Here g is aligned with the z axis and the centrifugal acceleration is aligned with the r axis. The three-dimensional numerical simulations used a Cartesian coordinate system.

Model and Numerical Methods

The model presented here attempts to produce a temperature field with perfectly paraboloidal isotherms. However, with constant material properties, a temperature field with paraboloidal isotherms is not achievable in the absence of convection because the temperature field is not a solution of equation 16 without internal heat generation. Paraboloidal isotherms would be represented by the equation:

$$T = bz + cr^2 + T_{ref} \tag{17}$$

whose gradient is:

$$\nabla T = b\hat{z} + 2cr\hat{r} \tag{18}$$

and Laplacian is:

$$\nabla^2 T = 4c \tag{19}$$

which does not satisfy the steady state energy conservation equation 16 in the absence of convection without the inclusion of a heat source. However, as will be seen, even without the heat source a *nearly* parabolic temperature field can be achieved and does demonstrate thermal stability in a rotating fluid on earth. A perfect paraboloidal temperature field could have been input into the model and the temperature not solved for. However, we felt that using the material properties of a common substance and solving for the temperature field would be more realistic.

The model and boundary conditions are shown on a half cross-section slice in figure 4. In nearly all the simulations, no heat source term was included. However, several cases are discussed in the results section where a heat source was included in the simulations. As discussed above, a nearly parabolic temperature field can be achieved without a heat source term. However, the inclusion of a heat source makes the isotherms more nearly parabolic. In the 3D simulations, a cylinder with paraboloidal ends was used. For the 3D simulations, the centerline shown in figure 4 was not a boundary. Because of symmetry in the 2D

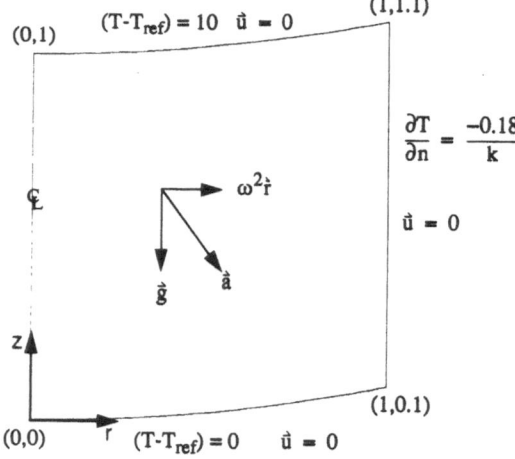

Figure 4. Model and boundary conditions used to investigate the thermally stable configuration for a rotating fluid with a constant background acceleration field.

axisymmetric simulations the centerline was a boundary and only 1/2 of the paraboloid was solved for. The upper and lower boundaries are parabolic as described by:

$$z = \frac{(T - T_{ref})}{10} + 0.1r^2 \tag{20}$$

where $(T - T_{ref}) = 0$ for the lower boundary and $(T - T_{ref}) = 10$ for the upper boundary. For reference, the corner point locations are labeled (r,z) in figure 4. Notice that the upper and lower boundaries have outward unit normals:

$$\hat{n} = \pm \left(\frac{\hat{z} - 0.2r\hat{r}}{\sqrt{1 + 0.04r^2}} \right) \tag{21}$$

The boundary conditions labeled in figure 4 are listed in table 1.

Table 1. Boundary conditions used in the model.

Boundary	Temperature ($^{\circ}$C) or Heat Flux (cal/s-cm^2)	Velocity (cm/s)
upper	$(T - T_{ref}) = 10$	$\hat{u} = 0$
lower	$(T - T_{ref}) = 0$	$\hat{u} = 0$
centerline (For 2D axisymmetric only)	$\frac{\partial T}{\partial n} = 0$	$u_r = 0$
outer	$\frac{\partial T}{\partial n} = \frac{-0.18}{k}$	$\hat{u} = 0$

Notice from table 1 that no-slip boundary conditions were applied to all external surfaces, resulting in an enclosed fluid. The angular component of velocity was not solved for in the 2D axisymmetric simulations because of the possibility that the flow could vary in the angular direction. To assume $\partial \hat{u}/\partial \theta = 0$ in this model *a priori* is probably an invalid assumption. Hence, the 2D model was used without Coriolis effects and given a zero velocity in the angular direction:

$$u_\theta = 0 \tag{22}$$

The outer boundary heat flux condition listed above is not coincidental. Assuming symmetry about the z axis, the gradient of temperature is:

$$\nabla T = \frac{\partial T}{\partial z}\hat{z} + \frac{\partial T}{\partial r}\hat{r} \tag{23}$$

From the model and equations 20 and 23, it would be expected that everywhere:

$$\frac{\partial T}{\partial z} \cong 10 \; ^{\circ}\text{C/cm} \tag{24}$$

23

Using equations 17 and 20 gives $c = -1$ $^{\circ}$C/cm^2. Using $r = 1$ cm at the outer boundary gives:

$$b = 10 \ ^{\circ}\text{C/cm} \qquad \text{and} \qquad 2cr = -2 \ ^{\circ}\text{C/cm} \tag{25}$$

Relating equations 18, 23 and 25 gives:

$$\frac{\partial T}{\partial r}\hat{r} = -2 \ ^{\circ}\text{C/cm} \tag{26}$$

which, with $k = 0.09$cal/cm-s-$^{\circ}$C, leads to the boundary condition:

$$-k\frac{\partial T}{\partial r}\hat{r} = -k\frac{\partial T}{\partial n}\hat{n} = q'' = 0.18 \ \text{cal/s-cm}^2 \tag{27}$$

where q'' is the heat flux out of the boundary. At the upper and lower boundaries the isotherms are almost perfectly paraboloidal because the mesh is almost perfectly paraboloidal. The isotherms near the outer boundary and centerline are approximately paraboloidal.

It may at first appear unusual to set $u_\theta = 0$ as a boundary condition. However, the model is in the rotating non-inertial frame of reference, wherein the solid boundaries are motionless. In the inertial (lab) frame of reference the solid boundaries rotate at rotation rate ω.

The conservation equations 14 - 16 in dimensional form were solved using a modified version of FIDAP 5.04, a finite element based code[6]. For most of the results presented here the fluid was assumed to be at steady state. Some transient simulations for selected rotation rates are also presented. In all the simulations a fixed-grid approach (nodal points spatially fixed) was used. The results presented hereafter were checked for convergence to within a specified absolute tolerance (1×10^{-6}) for both the normalized velocity and the residual error norms. Spatial convergence was ascertained by comparing the 3D results with very high resolution 2D results. The steady state simulations involved 6561 nodes using 4 node isoparametric quadrilateral elements for 2D simulations and 6015 nodes using 8 node isoparametric brick elements for the 3D simulations. The mesh for the 3D steady state results is shown on a corner cut in figure 5. The transient 3D simulations had slightly less spatial resolution, involving 4044 nodes using 8 node isoparametric brick elements.

The thermophysical properties of Ge, a common semiconductor, were used for the fluid and are listed in table 2.

Table 2. Properties of molten germanium at 950 $^{\circ}$C

Density[7,8] = 5.5 g/cm^3
Viscosity[7] = 7.4 x 10^{-3} g/cm-s
Specific Heat[9] = 0.091 cal/g-$^{\circ}$C
Thermal Conductivity[7] = 0.090 cal/cm-s-$^{\circ}$C
Thermal Expansion Coefficient[8] = 9 x 10^{-5} $^{\circ}$C^{-1}

The 2D axisymmetric simulations used a cylindrical coordinate system. The 3D simulations used a Cartesian coordinate system with rotation about the x axis, as shown in figure 5. As will be seen, the choice of coordinate systems has no bearing on the results, but a code check was made using a Cartesian coordinate system with rotation about the z axis.

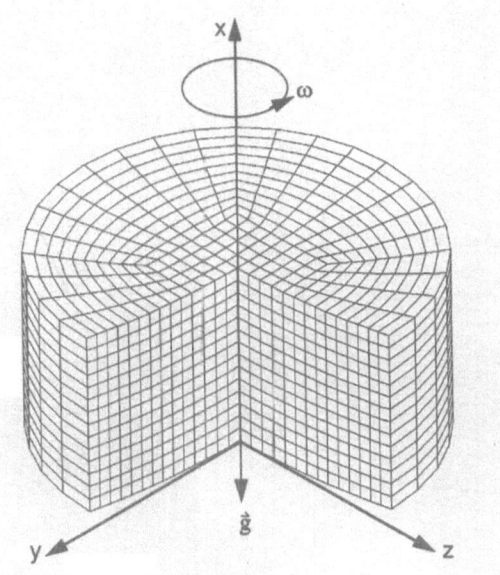

Figure 5. The 3D mesh using 8 node bricks of a cylinder with paraboloidal ends. Corner cut shown.

Results

The rotation rate at which convection will cease can be calculated *a priori* for perfectly paraboloidal isotherms. No convection should occur at the thermal stability rotation rate. At that rotation rate, the acceleration vector must be antiparallel to the gradient of the temperature at each and every point. Since the upper and lower boundaries are isothermal and paraboloidal, and the fact that the other two boundaries have boundary conditions corresponding to paraboloidal isotherms, the acceleration field and the gradient of the thermal field must have the same form, i.e.:

$$\vec{a} = (-981)\,\hat{z} + \omega^2 \hat{r} \tag{28}$$

$$\nabla T = 10\hat{z} - 2r\hat{r} \tag{29}$$

Thus, at the rotation rate for thermal stability there is nearly no convection and $\vec{a} \times \nabla T = 0$. Solving for the rotation rate gives:

$$\omega = \sqrt{\frac{2 \times 981}{10}} \cong 14 \text{ rad/s} \tag{30}$$

However, the temperature field is not perfectly paraboloidal everywhere without the inclusion of a heat source. Hence, the convection at the predicted thermal stability rotation rate will not be identically zero, but should be at a minimum.

The thermal field in the absence of convection is shown in figure 6. The isotherms are nearly paraboloidal. Because a low Prandtl number fluid was used in the model, the isotherms are nearly the same as in figure 6 when convection is present, with the parameters used in this study. At $\omega = 0$, the Rayleigh number based on a 1 cm radius and 1 °C radial temperature difference is about 365, which represents a reasonably small driving force.

Without the inclusion of the Coriolis effect, figures 7, 8 and 9 show the flow modes that

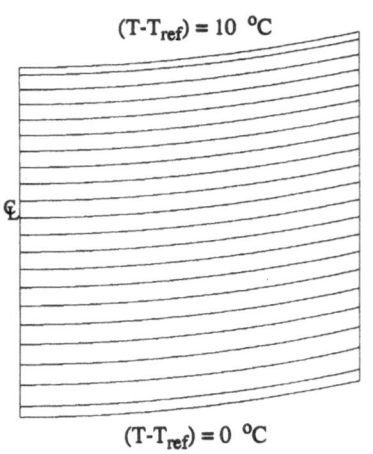

$(T-T_{ref}) = 10\ ^oC$

$(T-T_{ref}) = 0\ ^oC$

Figure 6. Calculated isotherms in the fluid in the absence of convection, 1/2 cross-section shown.

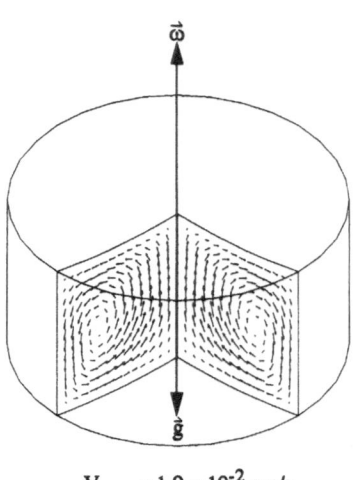

$V_{max} = 1.9 \times 10^{-2}$ cm/s

Figure 7. Fluid flow at $\omega = 13$, without Coriolis effect.

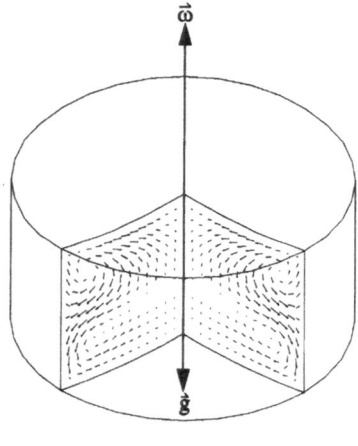

$V_{max} = 3.7 \times 10^{-3}$ cm/s

Figure 8. Fluid flow at $\omega = 14$, without Coriolis effect.

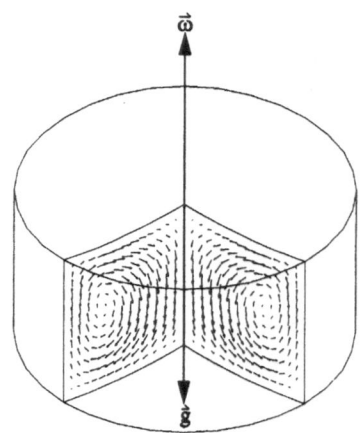

$V_{max} = 2.2 \times 10^{-2}$ cm/s

Figure 9. Fluid flow at $\omega = 15$, without Coriolis effect.

develop for rotation rates less than ($\omega = 13$), equal to ($\omega = 14$) and greater than ($\omega = 15$) the rotation rate calculated for thermal stability. The flow mode for $0 < \omega < 14$ is nearly identical to that in figure 7, and for $14 < \omega < 20$ nearly identical to that in figure 9. At the rotation rate at which thermal stability is predicted ($\omega = 14$), the flow modes for $\omega < 14$ and $\omega > 14$ seem to cancel one another and the flow nearly stops. This occurs because the buoyancy due to the centrifugal acceleration field equals the negative of the buoyancy due to the gravitational acceleration field at $\omega = 14$.

A plot of the maximum velocity demonstrates this nearly convectionless regime, as shown in figure 10. The results are presented both with and without the Coriolis effect in figure 10, and include 2D axisymmetric results without the Coriolis effect. The 2D results are so nearly identical to the 3D that the 2D curve lies on top of the 3D without Coriolis effect curve, except at $\omega = 14$ where the 2D convection is more suppressed, probably due to the high resolution in 2D. The velocities in the thermally stable configuration are non-zero, mainly due to the imperfect paraboloidal isotherms. However, the flow is one to two orders of magnitude less at $\omega = 14$ than at $\omega = 0$ or $\omega = 20$.

Inspection of equation 19 indicates that when a small heat generation term is applied to the fluid a further reduction in convection may occur due to the isotherms becoming more perfectly parabolic. Physically, this heat generation could result from Joule heating. Thus, relating equations 16 and 19 gives:

$$k\nabla^2 T = -q_s = k\,(4c) \tag{31}$$

Hence, with $c = -1\ ^\circ C/cm^2$ and $k = 0.090$ cal/cm-s-$^\circ C$ the heat generation per unit volume is:

$$q_s = 0.36\ cal/s\text{-}cm^3 \tag{32}$$

When this small heat generation was added in the 3D model, the maximum velocity at $\omega = 14$ increased by 1% from the case without a heat source. The lack of further convective suppression in the 3D model is attributed to mesh resolution. In the 2D axisymmetric simulation, the convection was further suppressed by about an order of magnitude, with $V_{max} = 7.1 \times 10^{-5}$ cm/s.

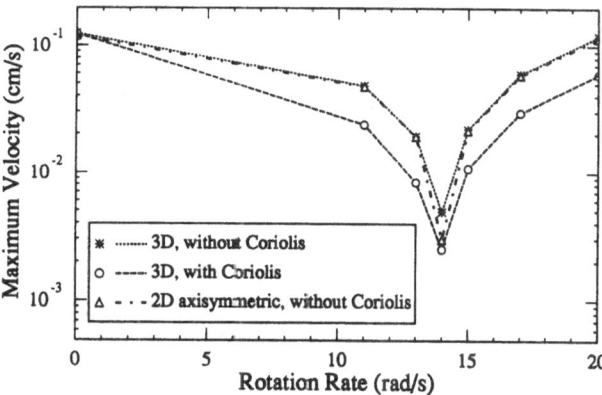

Figure 10. Maximum velocity in the fluid as a function of rotation rate both with and without the Coriolis force. Two-dimensional high resolution axisymmetric results are also shown.

If a more accurate value of the rotation rate calculated by equation 30, $\omega = \sqrt{196.2} \cong 14.00714104$, is used with the above heat generation a further reduction in convection would be expected to occur. However, no further reduction in convection occurred in the 3D simulations, and the maximum velocity increased by 1%. The convection began to increase in the 2D axisymmetric case also, with $V_{max} = 1.1 \times 10^{-4}$ cm/s, but it is still an order of magnitude less than at $\omega = 14$ without heat generation. From the above results, it can be concluded that this 3D model has reached its resolution limit, with a numerical "noise" level on the order of 2.5×10^{-3} cm/s for the velocity.

It should be mentioned that controlling the rotation rate to within a tolerance of ± 0.01 rad/s would be challenging experimentally. In addition, in an experiment the value of earth's gravity would have to be known to several significant figures at the location of the experiment to require a small tolerance on ω (g is known to vary between 978.039 and 983.217 cm/s^2 at sea level[10]). Thus, if this were an actual experiment there would always be some level of convection.

Transient simulations were done at rotation rates $\omega = 0$, $\omega = 14$ and $\omega = 20$ with the inclusion of Coriolis effects to test solution stability. A plot of the maximum velocity as a function of time for the three rotation rates is shown in figure 11. The flows reach the previously calculated steady state velocities in a relatively short time, typically less that a minute (0.0025 cm/s from the steady state analysis compared to 0.0026 cm/s from the transient analysis at $\omega = 14$). Also, the $\omega = 14$ case was done both with an initial zero velocity field and with the velocity field calculated at $\omega = 0$. Both of these resulted in the steady state flow. Hence, the steady state flow is not dependent on initial conditions. The flow patterns and velocities are identical within numerical error to those calculated above in the steady state simulations. The above transient simulations, which used the steady state velocity field at $\omega = 0$ as an initial condition, should not be construed as a true spin-up of the fluid, for they are not. A true spin-up transient simulation in the rotating frame of reference would have non-slip boundary conditions and an initial velocity field that included the effect of using the model at two different rotation rates. In addition, the zero initial velocity field where density gradients are present is aphysical. The above transient simulations used the two different initial velocity fields for the sole purpose of testing the temporal stability of the steady state simulations.

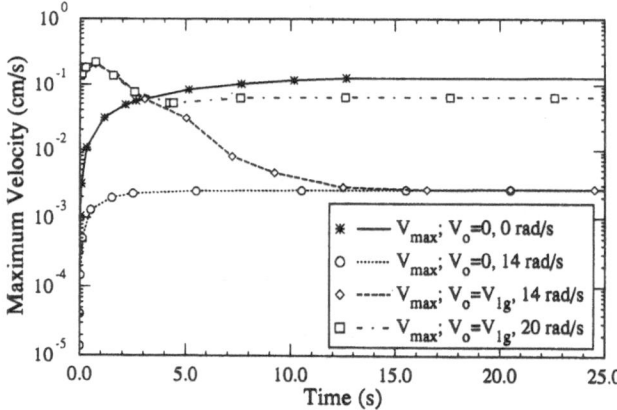

Figure 11. Maximum velocity in the fluid as a function of time for $\omega = 0$, $\omega = 14$ and $\omega = 20$ with Coriolis effect.

As a numerical check, a 3D pseudo-parallelepiped model with parabolic ends was examined. The mesh of the model is very different from the 3D cylinder and is shown in figure 12 (8125 nodes for $\omega = 14$; 6348 nodes for all other ω). The boundary conditions outlined for the 3D cylinder were applied to this model. The constant heat flux boundary condition applied to the outer walls did not necessarily have to produce nearly parabolic isotherms. However, it turned out that the isotherms were again nearly parabolic. Only the results with the inclusion of the Coriolis effect are presented.

Again at $\omega = 14$ a minimum in convection was encountered. In addition, the velocities parallel and perpendicular to the rotation vector were minima at $\omega = 14$. Figure 13 shows the flow pattern at $\omega = 13$ on two planes perpendicular to the axis of rotation. One plane is near the top at $z = 0.9$ cm and the other is near the bottom at $z = 0.3$ cm (z ranged from 0.0 to 1.2 cm). The fluid flow is primarily in a thin layer close to the walls. Near the cooler bottom end, the general flow is in the same sense as the rotation for $\omega < 14$ and counter to the rotation for $\omega > 14$. The opposite occurs at the hotter end, where the general flow is counter to the rotation for $\omega < 14$ and in the same sense as the rotation for $\omega > 14$. A plot of the maximum velocity and the maximum velocities parallel and perpendicular to the rotation vector demonstrates this regime of minimum convection, as shown in figure 14. The results in figure 14 include the Coriolis effect. Here again, the flow is about 2 orders of magnitude less at $\omega = 14$ than at $\omega = 0$ or $\omega = 20$. If geostrophic flow related phenomena[2] were occurring here, such as that predicted by the Taylor-Proudman theorem, the fluid flow parallel to the rotation vector would approach zero. Such phenomena might be expected at high rotation rates. However, for the model presented here it appears that suppression of convection occurs solely due to thermal stability, and there is no indication that geostrophic flow related phenomena are occurring.

The coordinate system used to solve the 3D pseudo-parallelepiped with parabolic ends model was rotated 90 degrees with respect to the 3D cylinder model such that rotation was

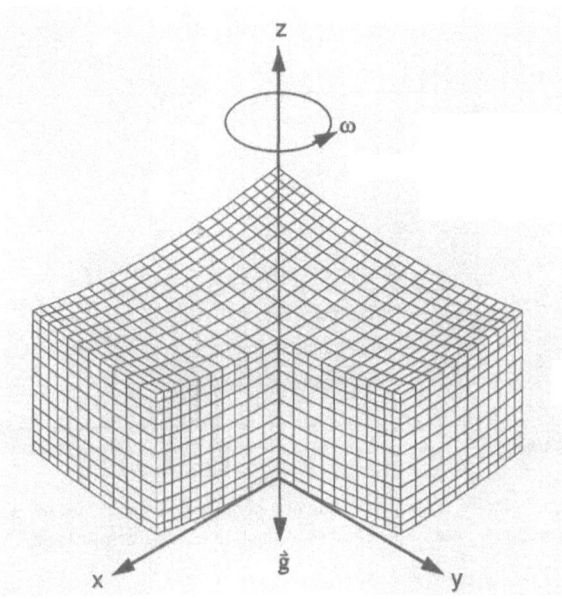

Figure 12. The three-dimensional mesh of a pseudo-parallelepiped with paraboloidal ends using 8 node bricks. Corner cut shown.

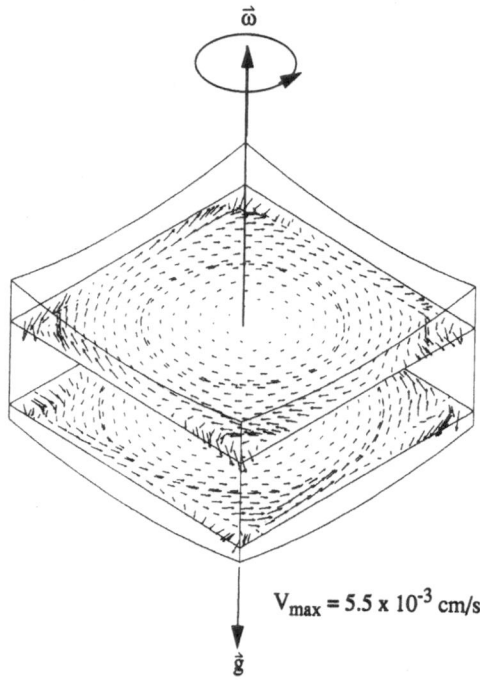

$V_{max} = 5.5 \times 10^{-3}$ cm/s

Figure 13. Fluid flow at $\omega = 13$, with Coriolis effect. The planes shown are perpendicular to the axis of rotation.

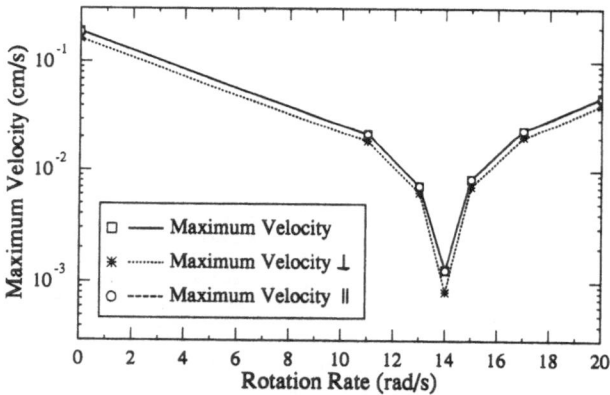

Figure 14. Maximum velocity, maximum velocity parallel to rotation vector and the maximum velocity perpendicular to the rotation vector in the fluid as a function of rotation rate (with Coriolis force).

about the z axis for the 3D pseudo-parallelepiped and about the x axis for the 3D cylinder. The good agreement between the results serves as an additional check on the numerical code.

CONCLUSIONS

The criteria for thermal stability is when the acceleration field is orthogonal to the isotherms at every point in the fluid, and the density gradient is parallel to the acceleration vector everywhere. For a rotating fluid on earth the rotation rate at which the convection is a minimum can be calculated once the equation describing the paraboloidal isotherms is given, or vice versa. Our numerical results show that there is at least one thermal field leading to thermal stability (a nearly convectionless regime) for a fluid rotating on earth. The isotherms that result in thermal stability in a rotating fluid on earth are parabolic, or very nearly so.

The Coriolis force does not alter the thermal stability configuration or the rotation rate at which the convection is minimized. Actually, convection was decreased by the inclusion of the Coriolis force, although the flow patterns were changed drastically. This was expected since the nondimensional ratio of buoyancy to Coriolis effect was small (Gr/Ta or Ra/Ta).

Transient analyses indicated that the flows were stable for rotation rates in the range of 0 to 20 rad/s. Changing the enclosure and the mesh did change the value of the maximum velocity. However, the minimum convection still occurred at the rotation rate where the acceleration vector was perpendicular to the isotherms at the top and bottom of the model. The suppression of convection at a particular rotation rate in a centrifuge on earth is independent of the geometry of enclosure as long as the isotherms are nearly parabolic.

INDUSTRIAL APPLICATIONS OF THE THERMAL STABILITY PHENOMENA

The work here suggests another method to suppress convection during the growth of semiconductor crystals by directional solidification,[11] as depicted in figure 15. The ampule

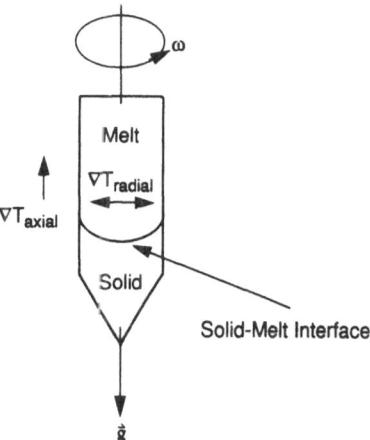

Figure 15. Vertical directional solidification and the thermal instabilities that induce convection. Cross-section of a cylindrical ampule is shown. Here ∇T_{radial} is the radial temperature gradient and ∇T_{axial} is the axial temperature gradient.

is rotated about its longitudinal axis. If the interface is concave and nearly paraboloidal, convection will decrease in the vicinity of the solid-melt interface as the rotation rate is increased because the net acceleration vector becomes more perpendicular to the interface, which is an isotherm. A minimum in convection will be achieved near the interface at some rotation rate. As the rotation rate is further increased, the convection will increase and reverse direction as the net acceleration vector again becomes non-perpendicular to the interface. At the minimum, convection may be sufficiently weak that it does not influence impurity segregation and uniform doping would be achieved.

This method has the advantage of not being dependent on material properties so long as the interface is concave toward the melt. In addition, the above method should be inexpensive to implement on existing crystal growth apparatuses.

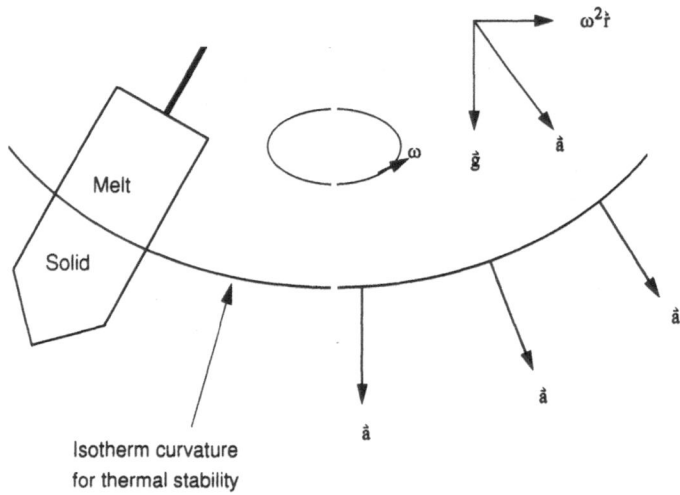

Figure 16. Ampule configuration when the melt-solid interface curvature is approximately equal to the thermally stable isotherm curvature during directional solidification in a centrifuge. (2D cross-section through the axis of rotation shown).

Directional solidification of bulk semiconductors in centrifuges may be another commercial application. Here, the solid-melt interface would have to have a shape approximately equal to a section of a paraboloid as depicted in figure 16. This method also has the advantage of increasing the hydrostatic pressure at the growth interface, which would reduce problems with gas bubbles.

ACKNOWLEDGMENTS

We are thankful to the Computational Material Science Laboratory at NASA Lewis Research Center for the use of their facilities and computer resources. We would also like to thank W. R. Wilcox of Clarkson University for his helpful advice.

NOMENCLATURE

Dimensional Quantities

a   Net acceleration; vector sum of gravitational and centrifugal accelerations, cm/s^2

C_p Specific heat, $cal/g\text{-}^oC$

$\hat{e}$ Unit directional

g   Gravitational acceleration, cm/s^2

k   Thermal conductivity, $cal/cm\text{-}s\text{-}^oC$

L   Characteristic length, cm

$\hat{n}$ Unit normal

p   Pressure, $g/cm\text{-}s^2$

P   Combined effect of local pressure and static centrifugal and gravitational forces ($P = p + \rho_o gh - \rho_o \omega^2 r^2/2$; where h is the distance opposite to gravity from any chosen reference plane[12] and r is the radial distance perpendicular to the rotation axis[12]), $g/cm\text{-}s^2$

q"   Heat flux, $cal/s\text{-}cm^2$

q_s Heat generation per unit volume, $cal/s\text{-}cm^3$

r,θ,z Cylindrical spatial coordinates, cm, rad, cm

$\vec{R}$ Position vector, cm

T   Temperature, ^oC

t   Time, s

$\vec{u}$ Vector velocity, cm/s

V   Velocity in the fluid, cm/s

x,y,z   Cartesian spatial coordinates, cm, cm, cm

Greek

β Thermal expansion coefficient, $^oC^{-1}$

μ Dynamic viscosity, g/cm-s

ρ Density, g/cm^3

ω Rotation rate, rad/s

Subscripts

axial   Axial value

C   Cold

H   Hot

L   Refers to a length

max   Maximum value

o   Initial of reference value

r,θ,z Refers to cylindrical spatial coordinates

radial   Radial value

ref   Reference value

1g   Value at the 1 g level

Nondimensional Quantities

Gr	Grashof number, $\rho^2 a\beta(T_H-T_C)_L L^3/\mu^2$
Pr	Prandtl number, $\mu C_p/k$
Ra	Rayleigh number, $\rho^2 C_p a\beta(T_H-T_C)_L L^3/\mu k$
Ta	Taylor number, $4\rho^2\omega^2 L^4/\mu^2$

Symbols and Diacritical Marks

$\tilde{\nabla}^2$	Vector Laplacian operator
$\perp$	Perpendicular to the rotation vector
$\parallel$	Parallel to the rotation vector
$\wedge$	Unit Vector
$\rightarrow$	Vector quantity

REFERENCES

1. D. H. Matthiesen, Ph.D. thesis, Massachusetts Institute of Technology, Cambridge, MA (1988).
2. D. J. Tritton, "Physical Fluid Dynamics," Second Edition, Oxford University Press, New York (1988).
3. W. Weber, G. Neumann and G. Muller, "Stabilizing Influence of the Coriolis Force During Melt Growth on a Centrifuge," *J. Crystal Growth,* 100: 145 (1990).
4. H. Rodot, L. L. Regel, and A. M. Turtchaninov, "Crystal Growth of IV-VI Semiconductors in a Centrifuge," *J. Crystal Growth,* 104: 280 (1990).
5. W. A. Arnold, Ph.D. thesis, Clarkson University, Potsdam, New York (1993).
6. M. Engelman, FIDAP Theoretical Manual, Fluid Dynamics International, Inc., 500 Davis Street, Suite 600, Evanston, Illinois 60201 (1990).
7. V. M. Glazov, S. N. Chizhevskaya and N. N. Glagoleva, "Liquid Semiconductors," Plenum Press, New York (1969).
8. T. Iida and R. I. L. Guthrie, "The Physical Properties of Liquid Metals," Oxford University Press, New York (1988).
9. Y.S. Touloukian and E. H. Buyco, "Thermophysical Properties of Matter," Vol. 4, IFI/Plenum Data Corp., New York (1970).
10. D. Haliday and R. Resnick, "Fundamentals of Physics," Second Edition, John Wiley and Sons Inc., New York: 278 (1986).
11. NASA Invention Disclosure, submitted by W. A. Arnold, NASA GSRP (1992).
12. R. Bird, W. Stewart and E. Lightfoot, "Transport Phenomena," John Wiley and Sons Inc., New York: 45 and 98 (1960).

CONVECTIVE FLOWS DURING CRYSTAL GROWTH IN A CENTRIFUGE

Vadim A. Urpin

A.F. Ioffe Institute of Physics and Technology
194201 St. Petersburg, Russia

ABSTRACT

Convective phenomena in the melt strongly influence crystal growth in a centrifuge. Large scale convective flows are induced, because hydrostatic equilibrium cannot be established in the melt, if the temperature gradient is not parallel to the "total" (gravitational plus centrifugal) acceleration. The convective velocity depends on parameters of the centrifuge and the furnace, as well as on the properties of the melt. A simple analytical expression for the convective velocity is obtained that includes the effect of both driving factors, the radial temperature gradient and the non-uniformity of the acceleration.

INTRODUCTION

Convective phenomena play an important role during solidification and strongly influence the properties of the resulting ingots (structure, distribution of impurities, etc.).[1] Convective phenomena in the melt depend strongly on gravity and, therefore, a wave of interest has been excited both for space experiments with low gravity and for experiments with centrifuges where the total (gravitational plus centrifugal) acceleration can be several times more than earth's gravity. The very low gravity level offered by space laboratories is ideal in the respect of damping of thermal convection, although not free from many non-trivial difficulties.[2] Experiments under centrifugation also show the importance of convection and, hence, of gravity for solidification process.[3,4]

The phenomena involved in solidification from the melt are rather complex. In crystal growth in a centrifuge the situation is further complicated by the addition of acceleration gradient and Coriolis force. Several time and length scales, spanning many orders of magnitude, are simultaneously responsible for setting the flow, thermal and solutal fields. Other complicating effects to be considered are time dependency and multidimensionality, which can substantially determine the configuration of the resulting flow in the melt. Physically incorrect results may be produced by centrifugal crystal growth models, that do

Materials Processing in High Gravity, Edited by L.L. Regel
and W.R. Wilcox, Plenum Press, New York, 1994

not take into account such factors as the presence of a radial temperature gradient, non-uniformity of the acceleration, etc.[5]

The aspect ratio of the melt at the beginning of solidification is usually sufficiently large, so that the quasi steady-state solidification conditions can be reached. However, toward the end of the growth, the aspect ratio becomes small. Most theoretical analyses, including the present paper, choose a conveniently large aspect ratio to allow the quasi steady-state assumption. This assumption is unacceptable for short aspect-ratio melts where the temperature and velocity fields are dependent on the boundary conditions at the melted end. Such a solidification process never reaches a steady state and will not be considered in the present paper.

In analysis of solidification in a centrifuge an important ingredient, which must be included, is the inherent three-dimensionality of the convective flow resulting from the three-dimensionality of the driving mechanisms. In centrifugal crystal growth some flow regimes may be driven by both radial and axial thermal gradients in the fluid. The Coriolis force, which essentially determines the convective regime in the melt,[6] is generally not axisymmetric. Therefore, the convective flows, which are driven by the Coriolis force, should be three-dimensional. On the contrary, convection in the regimes with non-dominating Coriolis force can be axisymmetric.

The resulting velocity and temperature fields driven by all the above factors may be rather complicated. The present paper considers some aspects of hydrodynamic processes in the melt. We analyze mainly the most general characteristics of flows in a rotating fluid and their dependence on the parameters of real experiments in centrifuges.

HYDROSTATIC EQUILIBRIUM

The equation of hydrostatic equilibrium in a rotating fluid can be written in the form:

$$-\frac{\nabla p}{\rho} + \vec{g} = 0 \quad , \quad \vec{g} = \nabla\left(\psi + \frac{1}{2}\Omega^2 s^2\right) \quad , \tag{1}$$

where $\vec{\Omega}$ is the angular velocity; p and ρ are the pressure and density, respectively; $\vec{g}$ is the total (gravitational plus centrifugal) acceleration; ψ is the gravitational potential; $\nabla\psi = \vec{g}_0$, with $\vec{g}_0$ being the standard earth's gravity; and s is the distance from the rotational axis. Equation (1) implies the lack of any hydrodynamic motion in the fluid with the exception of rotation with angular velocity $\vec{\Omega}$. Calculating the curl of Eq. (1), one gets:

$$\vec{g} \times \nabla\rho = 0 \quad . \tag{2}$$

In the Boussinesq approximation, which is rather accurate for laboratory experiments with liquids, we have:

$$\rho = \rho(T) = \rho_0 - \rho_0\beta(T - T_0) \quad , \tag{3}$$

where ρ_0 is the constant mean density, T_0 is the average temperature, and β is the thermal expansion coefficient. Therefore, assuming β independent of T:

$$\nabla\rho = -\rho_0\beta\nabla T \tag{4}$$

Substituting expression [Eq. (4)] into Eq. (2), we obtain the necessary condition for hydrostatic equilibrium in a nonuniformity heated rotating fluid:

$$\nabla T \times \vec{g} = 0 \quad, \tag{5}$$

i.e. the temperature gradient ∇T must be parallel to the total acceleration g. It appears that if condition [Eq. (5)] is not fulfilled in any part of the melt, hydrostatic equilibrium cannot be established. In real experimental conditions, it is nearly impossible to satisfy condition [Eq. (5)] everywhere within the melt. For hydrostatic equilibrium, one would need a specially designed furnace producing $\nabla T \parallel \vec{g}$ at every point in the ampoule. Thus convection in the melt is unavoidable during solidification in a centrifuge. However, under some conditions the convection may be so weak near the freezing interface, that the doping concentration profile corresponds to that expected in the absence of convection.

CONVECTIVE FLOW IN THE MELT

The set of equations governing the velocity and temperature distributions in the melt in the Boussinesq approximation are:[7]

$$\dot{\vec{v}} + (\vec{v}\cdot\nabla)\vec{v} + 2\vec{\Omega}\times\vec{v} = -\frac{\nabla p'}{\rho_0} - \vec{g}\beta T' + \nu\Delta\vec{v} \quad, \tag{6}$$

$$\dot{T}' + \vec{v}\cdot\nabla T = \kappa\Delta T' \quad, \tag{7}$$

$$\nabla\cdot\vec{v} = 0 \quad, \tag{8}$$

where $\vec{v}$ is the hydrodynamic velocity, $T' = T - T_0$ and p' are the variations of the temperature and pressure, respectively, from the mean values; $\vec{g} = \vec{g}_0 + \Omega^2\vec{s}$, $\vec{s}$ is the distance from the rotational axis; ν and κ are the kinematic viscosity and thermal diffusivity, respectively. The dot above a quantity denotes the partial derivative with respect to time. In what follows, we will use cylindrical coordinates (r, ϕ, z) associated with the axis of the ampoule.

If the aspect ratio is large and the quasi steady state condition is fulfilled, the velocity and temperature are given by:

$$(\vec{v}\cdot\nabla)\vec{v} + 2\vec{\Omega}\times\vec{v} = -\frac{\nabla p'}{\rho_0} - \vec{g}\beta T' + \nu\Delta\vec{v} \quad, \tag{9}$$

$$\vec{v}\cdot\nabla T = \kappa\Delta T' \quad, \tag{10}$$

$$\nabla \cdot \vec{v} = 0 \quad . \tag{11}$$

Both simulations and experimental data indicate the order of magnitude for the maximal value of the convective velocity, $v_m \sim 0.1$–0.2 cm/s. Since usually the axial temperature gradient is higher than the parallel one, the ratio of the convective and conductive heat fluxes is $\leq v_m R/\kappa$, where R is the radius of the ampoule. In many experiments $R \leq 1$ cm and $\kappa \geq 0.1$ cm^2/s, so that the ratio vR/κ can reach the order of 1 only in regions of the maximal velocity. It means that practically everywhere in the melt the thermal field is determined almost entirely by heat conduction and, at least in the zeroth approximation, one can determine the temperature distribution, solving the equation

$$\Delta T' \simeq 0 \tag{12}$$

with appropriate boundary conditions.

An estimate of the Reynolds number $Re = Rv/\nu$ for the above parameters and $\nu \sim 10^{-3}$ cm^2/s gives the value $\sim 10^2$. For typical values of ν, we have $Re > 1$ if $v > 0$ μm/s. Because of high Re numbers, the flow in the bulk melt can be estimated by neglecting viscosity. Viscosity is important at the boundary layer near the ampoule walls and the melting interface. It also appears that at flow velocities ≤ 0.1–0.2 cm/s, the nonlinear term in the Navier–Stokes Eq. (9) is less important than the Coriolis force. One can estimate that at $\Omega \sim 1$ s^{-1} with the above v_m and R, the inertial force in Eq. (9) is at least several times smaller than the Coriolis force. Therefore, in the dominant portion of the melt volume (with the exception of the boundary layers), the velocity is approximately given by:

$$2\vec{\Omega} \times \vec{v} = -\frac{\nabla p'}{\rho_0} - \vec{g}\beta T' \quad . \tag{13}$$

Calculating the curl of this equation and making use of Eq. (11), one gets:

$$2(\vec{\Omega} \cdot \nabla)\vec{v} = -\beta \vec{g} \times \nabla T \quad . \tag{14}$$

From this equation, we have the order of magnitude estimation:

$$v \sim \frac{\beta r}{2\Omega_r} |\vec{g} \times \nabla T| \quad , \tag{15}$$

where Ω_r is the radial component of $\vec{\Omega}$. Simple geometrical consideration provides $\Omega_r \sim \Omega(\Omega^2 s/g)$. Introducing the "gravitational" parameter $N = g/g_0$, we get $\Omega_r \sim \Omega\sqrt{N^2-1}/N$. Hence,

$$v \sim \frac{\beta r}{2\Omega_r} \frac{N}{\sqrt{N^2-1}} |\vec{g} \times \nabla T| \quad . \tag{16}$$

One cannot apply Eq. (15) at $N \simeq 1$, because in this case rotation is very slow and the Coriolis force does not dominate in Eq. (9).

As was mentioned above, the hydrodynamic velocity is nonzero if $\vec{g} \times \nabla T \neq 0$. We can write $\vec{g}$ and ∇T as $\vec{g} = \vec{g}_{\parallel} + \vec{g}_{\perp}$ and $\nabla T = \nabla_{\parallel} T + \nabla_{\perp} T$, where the indices $\parallel$ and $\perp$ mark the components of vectors parallel and perpendicular to the axis of the ampoule. Then,

$$\vec{g} = \nabla T = \vec{g}_{\parallel} \times \nabla_{\perp} T + \vec{g}_{\perp} \times \nabla_{\parallel} T + \vec{g}_{\perp} \times \nabla_{\perp} T \ . \qquad (17)$$

In laboratory conditions $|\nabla_{\perp} T|$ is normally smaller than $|\nabla_{\parallel} T|$. Therefore, the last term in the right−hand side of Eq. (17) can be omitted. The components of gravity are $g_{\parallel} \simeq g$ and $g_{\perp} \sim g(R/D)$, where D is the length of the centrifuge arm. The components of the temperature gradient can be estimated as:

$$|\nabla_{\parallel} T| \sim \Delta_{\parallel} T / L \ , \qquad \Delta_{\parallel} T = T_m - T_s \ , \qquad (18)$$

where T_s and T_m are the temperatures at the top of the ampoule and at the freezing interface, respectively, L is the length of the ampoule, and:

$$|\nabla_{\perp} T| \sim \Delta_{\perp} T / R \ , \qquad (19)$$

where $\Delta_{\perp} T$ is the temperature difference between the axis and the wall at a given z. Substituting this expression into Eq. (17), we obtain:

$$|\vec{g} \times \nabla T| \sim g \frac{\Delta_{\perp} T}{R}(1 + Q) \ , \qquad Q = \frac{R^2}{LD} \cdot \frac{\Delta_{\parallel} T}{\Delta_{\perp} T} \ . \qquad (20)$$

The parameter Q characterizes the relative contribution of the nonuniform gravity and radial temperature gradient to the generation of convection. If $Q < 1$, convection is mainly determined by the radial temperature gradient. On the contrary, if $Q > 1$ the acceleration gradient governs convection in the melt. The value of Q increases as the radial temperature gradient and the centrifuge arm length decrease.

Using expression (20), one can estimate the velocity from Eq. (16),

$$v \sim \frac{\beta g_0}{2 \Omega} \cdot \Delta_{\perp} T \cdot \frac{N^2}{\sqrt{N^2 - 1}} \cdot (1 + Q) \ . \qquad (21)$$

Since $\Omega \simeq \sqrt{g_0/D}(N^2 - 1)^{1/4}$, the convective velocity can be expressed also in terms of the centrifuge arm length and the gravitational parameter N instead of Ω,

$$v \sim \frac{1}{2} \beta \sqrt{g_0 D} \, \Delta_{\perp} T \cdot \frac{N^2}{(N^2 - 1)^{3/4}} \cdot (1 + Q) \ . \qquad (22)$$

It follows from Eqs. (21) and (22) that the convective velocity is determined by the thermal expansion coefficient, and the parameters of the furnace and centrifuge. It should be noted that even at a given gravity level N, convection depends on the parameters of the centrifuge

[Ω or D, see Eqs. (21) or (22)]. This dependence is due to the fact that convection is driven by two rotational forces, centrifugal and Coriolis, which cannot be described by only one parameter N. In principle, if convection is important for crystal growth, one can obtain crystals with different properties by doing experiments with the identical material, furnace, and gravity levels, but with different arms length of the centrifuge.

Evidently the furnace is very critical for convection. Convection may be much slower for a furnace with small $\nabla_{\parallel}T$ and $\nabla_{\perp}T$. However, even a furnace that gives a very small $\nabla_{\perp}T$, the axial gradient $\nabla_{\parallel}T$ should be nonzero to provide crystal growth. In this case, convection does not disappear although the convective velocity may be rather low,

$$v \sim \frac{\beta}{2} \cdot \sqrt{\frac{g_0}{D}} \cdot \frac{R^2}{L} \cdot \Delta_{\parallel}T \cdot \frac{N^2}{\left(N^2 - 1\right)^{3/4}} \quad . \tag{23}$$

For such a furnace, convection is driven by the acceleration gradient and becomes slower with an increase of the centrifuge arm length.

In a furnace producing strong parallel and perpendicular temperature gradients, a dependence of the convective velocity on the centrifuge arm length is more complicated. Since $Q \propto 1/D$, we have:

$$v \propto \left(\sqrt{D} + \frac{\alpha}{\sqrt{D}}\right) \quad , \tag{24}$$

where $\alpha = (R^2/L)(\Delta_{\parallel}T/\Delta_{\perp}T)$. At given parameters of material and furnace, v reaches its minimal value v_{min} at $D = \alpha$. For instance, if $R = 1$ cm, $L = 4$ cm, $\Delta_{\parallel}T = 30$ K, and $\Delta_{\perp}T = 0.1$ K, then $D = 75$ cm. An increase of the arm results in a slow ($\propto \sqrt{D}$) increase of the convective velocity.

Convection also depends on the gravity level N. Since the parameter Q is independent of acceleration, the convective velocity for a given centrifuge arm is proportional to $N^2/(N^2 - 1)^{3/4}$. This function decreases rather rapidly if N increases in the range $1 < N \leq 2$, reaching a minimal value at $N = 2$. At higher N this function increases slowly ($\propto \sqrt{N}$ at $N \gg 1$). For instance, if N increases from 2 to 12, the convective velocity increases only about twice. The order of magnitude estimation of the convective velocity gives, according to Eq. (22), at $Q < 1$ and $N = 2$:

$$v \sim 1.2 \times 10^3 \Delta_{\perp}T \ \mu m/s \tag{25}$$

(we assume $\beta \sim 10^{-4}$ 1/K and $D = 18$ m). Since usually $\Delta_{\perp} \sim 0.1$–1 K, the flow velocity may be $\sim 10^2$–10^3 μm/s.

CONCLUSION

In conclusion, we mention briefly the main results of the present study.

(1) Hydrostatic equilibrium during centrifuge crystal growth can be established only if the temperature gradient is parallel to the total acceleration at every point in the melt. The design of a furnace, which can produce such heating, is practically impossible. Therefore, in real experiments, hydrostatic equilibrium cannot be achieved and convection in the melt is inescapable.

(2) The main factors responsible for convection in a centrifuge crystal growth are the radial temperature gradient and the nonuniformity of the acceleration. The relative contribution of these factors is characterized by the parameter Q [see Eq. (20)]. If $Q < 1$, the radial temperature gradient contributes mainly to driving the convection. If $Q > 1$, the dominating factor is the acceleration gradient.

(3) The convective velocity in the melt depends not only the properties of the material, the parameters of the furnace, and the total gravity $N g_0$, but also on the centrifuge arm length. This dependence is due to the fact that the Coriolis force cannot be described in terms of the parameter N alone. In principle, it implies that crystals grown of the same material in identical furnaces at the same gravity level, but in different centrifuges, may have different properties.

(4) The characteristic value of the convective velocity v may be rather high, $v \sim 10^2$–10^3 μm/s. At low centrifugation, v decreases rather rapidly reaching its minimal value at $N = 2$. At higher values of N, the convective velocity amplifies slowly. An increase of N from 2 to 12 results in an increase of v only by the factor ~ 2.

REFERENCES

1. J.J. Favier, Recent advances in Bridgman growth modelling, *J. Cryst. Growth* 99:18 (1990).
2. L.L. Regel. "Material Processing in Space," Plenum Press, New York (1990).
3. H. Rodot, L.L. Regel, and A.M. Turchaninov, Crystal growth of IV–VI semiconductors in a centrifuge, *J. Cryst. Growth* 104:280 (1990).
4. G. Müller, *J. Cryst. Growth* 99:1242 (1990).
5. W.A. Arnold, W.R. Wilcox, F. Carlson, A. Chait, and L.L. Regel, Transport modes during crystal growth in a centrifuge, *J. Cryst. Growth* 119:24 (1992).
6. W. Weber, G. Neumann, and G. Müller, Stabilizing influence of the Coriolis force during melt growth on a centrifuge, *J. Cryst. Growth* 100:145 (1990).
7. L.D. Landau and E.M. Lifshitz. "Hydrodynamics," Nauka, Moscow (1986).

REMOVAL OF CONVECTIVE INSTABILITIES
IN LIQUID METALS BY CENTRIFUGATION

Alain Chevy,[1] Pascal Williams,[1] Michel Rodot[2] and Gérard Labrosse[3]

[1]Physique des Milieux Condensés
 Université Pierre et Marie Curie
 CNRS URA 782, T13 E4 B77
 4 place Jussieu
 75252 Paris Cedex 05, France
[2]Equipe Technologie de la Croissance Cristalline, CNRS
 1 place Aristide Briand
 92195 Meudon, France
[3]L.I.M.S.I. Université Paris–Sud
 BP 133
 91403 Orsay Cedex, France

ABSTRACT

Convective instabilities generated in liquid cells, under earth's gravity, for a destabilizing thermal gradient, can be removed by centrifugation. Experiments on molten tin show a reversible transition from an unstable to a stable regime, at a well defined value of the rotation rate. These observations can be understood by considering that the axial buoyancy force depends on both the density gradient and the acceleration generated along the axis of a fluid cell installed on a centrifuge.

INTRODUCTION

Recently there has been an increasing interest in the possibilities of producing advanced materials on centrifuges, as witnessed by the first workshop dedicated to "Materials Processing in High Gravity" held in the USSR.[1] The acceleration field produced by a centrifuge can be employed to manipulate the convection in the liquid or gaseous phases during solidification. This technique holds much promise for the growth of single crystals of improved quality and for the production of complex alloys. Centrifugal fluid physics provides an intriguing domain for study. Enhanced and nonhomogeneous body forces and the Coriolis effect have to be considered, especially in relation to their interactions with the thermal and thermal–solutal convections.

During crystal growth from the melt it is generally accepted that the convective regime is critical in determining the quality of the final crystal with respect to both structural defects and dopant distribution. Traditionally several methods have been proposed to control the convective regimes, such as enhanced control of the thermal environment, forced convection, and magnetic fields. More recently, Weber et al.[2] have illustrated that centrifugation is effective in transforming an unstable convective regime into a stable state. In the present work another case of stabilization of a liquid by centrifugation is presented.

Crystal growth methods can be characterized by the direction of the temperature gradient with respect to the earth's gravity vector. In the most widely used industrial technique, Czochralski, the temperature gradient has a similar direction to the gravity vector at the solid–liquid interface and is, therefore, thermally destabilizing, producing strong buoyancy convection and convective instabilities resulting in possible crystal defects. This case is similar to that of Weber et al.[2] who employed the inverted Bridgman method, with crystal growth starting from the top. They showed it was possible to suppress the temperature fluctuations, and resulting striations, during the growth of Te doped InSb crystals and invoked the Coriolis force during centrifugation to explain these interesting results. In the normal Bridgman method, the externally imposed temperature gradient is antiparallel to the earth's gravity, and thus thermally stabilizing, encouraging the formation of a stable convective regime.

The present work is a fluid experiment (without any compositional effect) where the influence of centrifugation on a thermally destabilizing system is considered. In contrast with the stabilizing case,[3,4] here centrifugation turns out to play an important role since it is observed that a nonstationary state under earth's gravity becomes stationary, when the rotation rate exceeds some well-defined value. This transition is reversible, which is the first indication that such a behavior is not due to the same mechanism as the one reported by Weber.[2] A simple interpretation is suggested.

EXPERIMENTAL

The furnace (Fig. 1) consisted of eight independent temperature zones, permitting precise control of the external ampoule thermal environment. Due to the hostile environment of the centrifuge a programmable computer was installed for temperature regulation and monitoring of the various experimental parameters.

Two distinctive types of quartz ampoules were employed. In the first type (type 1), quartz fingers descended into the melt, permitting the exploration of the thermal field by displacement of thermocouples with a remotely controlled motorized chariot. The quartz fingers also served to protect the thermocouples at elevated melt temperatures and determine their radial position. In the second type (type 2) of ampoule, thermocouples were held at a fixed height and in direct contact with the melt, suitable for low temperature investigations. The direct contact enhanced the sensitivity of the thermocouples to changes in melt temperature. In the case of the thermally destabilizing condition, ampoules of various diameters were employed to determine the influence of the melt diameter on the behavior of the fluid convection during centrifugation. The ampoules were sealed after evacuation to 10^{-6} mbar.

Both stabilizing and destabilizing furnace temperature gradients in the range 20 to 40 °C cm^{-1} were imposed. This is comparable to the imposed stabilizing gradient of 25 °C cm^{-1} during the solidification of PbTe ingots by Rodot et al.,[5] and the measured furnace gradient of 20 °C cm^{-1} during the solidification of Ge ingots by Chevy.[6] Melt temperatures were held at around 400 °C during experiments on tin in the destabilizing condition. The thermal environment outside the ampoules was monitored with thermocouples placed at the ampoule outer surface, at the furnace mouth and on the outer jacket.

Experiments were mounted on the basket of the "Laboratorie Central des Ponts et Chaussées" centrifuge situated at Nantes. There is a pivot at the end of the centrifuge arm which allows the basket vertical to align with the mean resultant acceleration vector (Fig. 2). Thus at low accelerations the radius varies significantly due to the swivel of the experimental basket. The fixed arm radius is 3.965 m and the distance from the end arm pivot to the basket floor is 1.535 m. This centrifuge, of maximum radius 5.5 m, can take a maximum mass of 2 kg up to 100 g, where g is earth's gravity. Transitions in rotation speed take 1 to 5 minutes in the range 1 to 10 g.

An important feature of the centrifuge environment, neglected in previous studies, is the forced convective cooling of the furnace. Preliminary experiments, on a tin melt at 700 °C, revealed a 24 °C drop of a thermocouple attached to the mouth of the furnace and

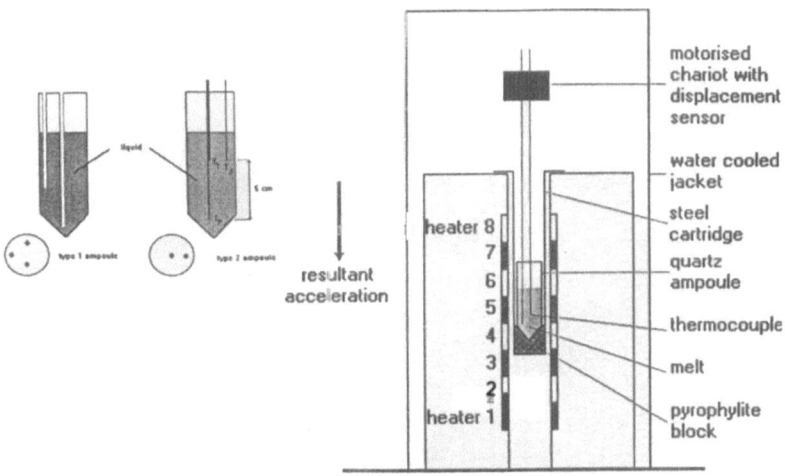

Figure 1. Multizone furnace with thermal mapping system and water–cooled envelope.

a 8 °C drop inside the furnace tube due to a 2 g acceleration. Thus, the external air currents have an important influence on the thermal environment of the ampoule, making interpretation of results difficult. Thus, the furnace was enclosed in an envelope, maintained at constant temperature by circulating water, to reduce these spurious effects.

RESULTS

We first present results obtained with a type 2 ampoule. The temperature gradient is defined as the difference between the upper and the lower thermocouple measurements divided by their separation. The tin melt had a height, $h = 80$ mm. The temperature gradient varied with centrifugal acceleration, as shown by Fig. 3. The changes in centrifuge speed are denoted by the vertical dashed lines, between which the centrifugal acceleration

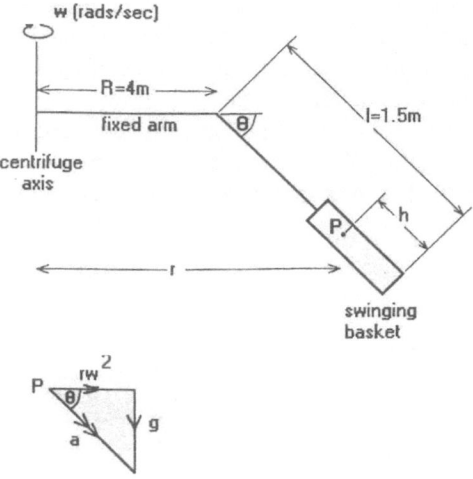

Figure 2. Geometry of the Nantes centrifuge and resultant acceleration.

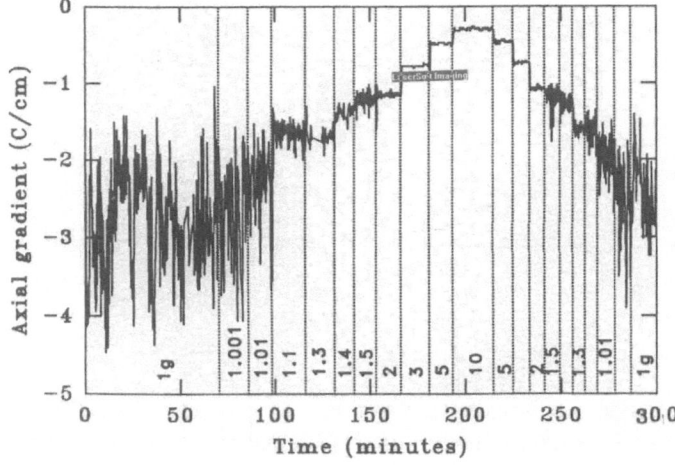

Figure 3. Axial temperature gradient measured inside destabilizing tin melt, height 80 mm and diameter 27 mm. Change of rotation speed occurs at vertical dashed lines, with the values between representing the resultant acceleration in g.

was constant. In normal laboratory conditions, at 1 g, the melt axial temperature gradient fluctuates in the range -2 to -4 °C cm^{-1}, indicating turbulent convection. The melt gradient is much smaller than the imposed gradient. The amplitude of the fluctuations is significantly reduced at a surprisingly small acceleration of 1.1 g. If the temperature resolution of 0.1 °C is taken into account, no fluctuations are observed at 3 g and above. A transition from a turbulent to a steady state convection regime occurs between 2 and 3 g. Upon reducing the centrifuge speed, this transition is seen to be perfectly reversible.

Often spikes are observed at the change of centrifuge speed, indicating that the molten metal is disturbed by the tangential acceleration.

There is an important reduction in the magnitude of the estimated average axial melt gradient with acceleration, as seen in Fig. 4, from 3.2 °C cm^{-1} at 1 g to 0.3 °C cm^{-1} at 10 g. The circles represent measurements made with decreasing acceleration and the squares increasing acceleration is observed before and after the turbulent – steady state transition. This reduction in melt gradient could be explained by the increased convection, resulting from the enhanced buoyancy, transporting the heat more effectively.

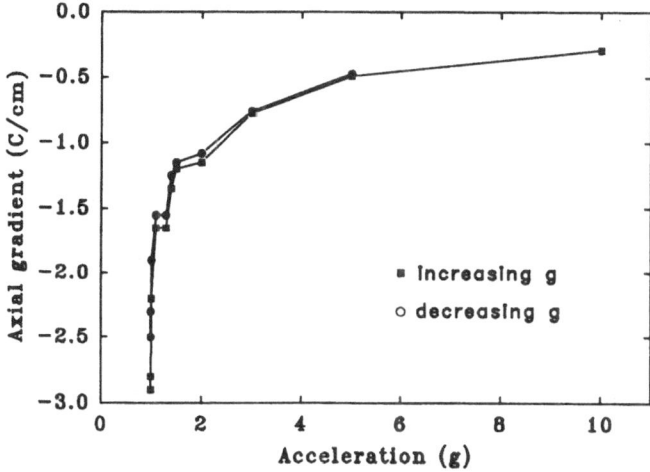

Figure 4. Axial temperature gradient measured inside destabilizing tin melt, height 80 mm and diameter 27 mm, as a function of resultant acceleration.

The variation of the radial temperature gradient with time, for the same experience, is given in Fig. 5. In normal laboratory conditions, this radial gradient is more than 10 times smaller than the axial gradient. When the resultant acceleration increases, the radial gradient shows the same behavior as the axial gradient, with an important reduction in fluctuations at 1.1 g, and the transition to a steady state between 2 and 3 g. The evolution of the radial gradient with acceleration is less marked than that of the axial gradient.

This experiment was repeated with the type 1 ampoule for a tin melt of height 70 mm and diameter 17 mm (the aspect ratio is 4.1 instead of 2.9). As in the previous experience, the melt axial gradient fluctuates around -3 °C cm^{-1} at 1 g, and there is a marked reduction in melt gradient with acceleration (Fig. 6). However, the turbulent to steady state transition occurs at 1.1 g. (Note the existence of an apparently meta–stable state at 1.005 g.)

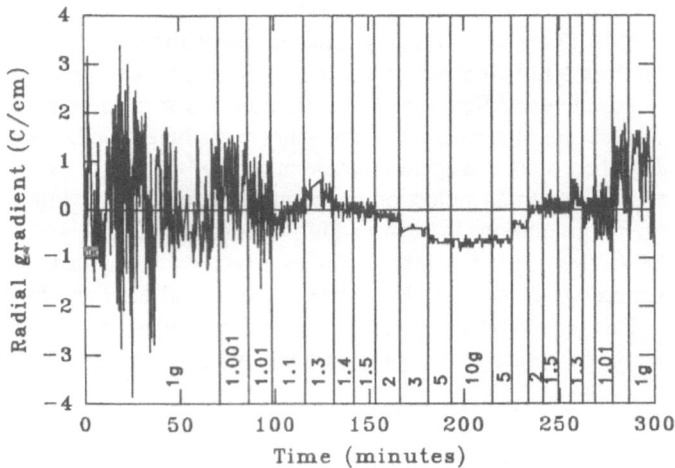

Figure 5. Radial temperature gradient measured inside destabilizing tin melt, height 80 mm and diameter 27 mm. Change of rotation speed occurs at vertical dashed lines, with the values between representing the resultant acceleration in g.

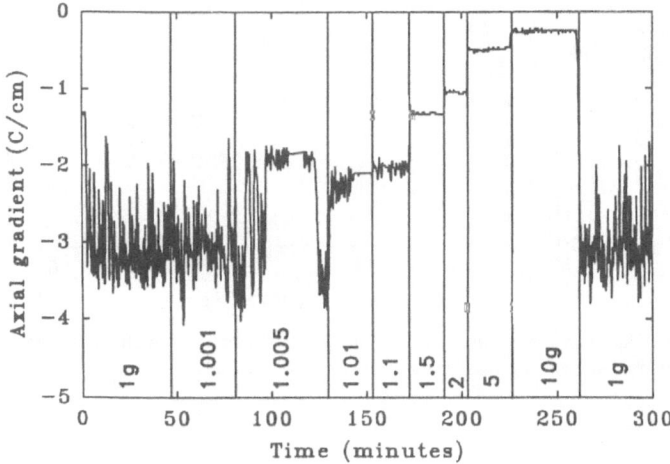

Figure 6. Axial temperature gradient measured inside destabilizing tin melt, height 70 mm and diameter 17 mm. Change of rotation speed occurs at vertical dashed lines, with the values between representing the resultant acceleration in g.

Comparison of these two experiments both with similar melt heights, but different diameters, leads to the conclusion that the required acceleration for the turbulent to stable transition increases when the aspect ratio decreases.

For the thermally stabilizing configuration, an extensive series of experiments was concluded with Sn, 70Sb–30Bi alloy, pure Ge and Al, and B and In doped Ge melts, held at fixed temperatures and during solidification with the vertical gradient freeze method. In all cases, melt temperatures remained steady during both normal laboratory conditions and centrifugation, and melt gradients were independent of acceleration. The stable convection regime was independent of centrifugal acceleration. This result agrees with the numerical analyses of Arnold et al.[3] and Fikri et al.[4] who found only slight influences of centrifugation on the thermal convection in the stabilizing vertical gradient freeze method.

INTERPRETATION

The coupling between rotation and buoyancy has motivated many studies since the first theoretical one by Chandrasekhar,[7] who showed the stabilizing influence of rotation (through the Coriolis term) on a thermally destabilized horizontal fluid layer submitted to earth's gravity field. Most of the analyses reported up to now were dedicated to the axisymmetrical centrifugation configuration, i.e. the rotation axis coincides with the symmetry axis of the fluid container. To present a survey of the large variety of transitions already observed is out of the scope of this contribution, as is a detailed analysis of the whole dynamics involved in our experiments. Rather, we shall limit ourselves to point out the presence of a stabilizing effect, specific to the "nonaxisymmetric" configuration, presumably responsible for the transitions reported herein. An appropriate Rayleigh number can thus be defined whose value decreases as the rotation angular speed, Ω, increases from zero.

Henceforth, the D (S) case will refer to the thermally destabilized (stabilized, respectively) configuration, obtained by imposing from outside, an axial component of the fluid thermal gradient parallel (antiparallel, respectively) to the resulting gravity, $\vec{g}_0 = - g_0 \cdot \hat{e}$ with $g_0 = g/(\sin \theta)$, g being earth's gravity. The external heating induces unavoidably, a radial component to the fluid thermal gradient, which triggers, in the S case, the convection. The S case is now well known, from experiments and numerical modeling. No unsteady convective regimes have been observed, and the Coriolis force acts merely to deflect the thermoconvective flow. The essential difference between the S and D cases comes, therefore, from the axial buoyancy contribution.

The following simple model shows that "nonaxisymmetric" centrifugation overstabilizes the S case and is able to stabilize the D case.

Let us consider a situation where the body force, $\rho\vec{g}$, is an axial buoyancy field, $\vec{g}(z) = - g(z) \cdot \hat{e}_z$, depending only on the axial coordinate, and z, $\rho(z)$, and g(z) being close to their mean values ρ_0 and g_0, respectively. Such a buoyancy field is stabilizing if $(d/(dz))(\rho g) < 0$, otherwise it is destabilizing.

Centrifugation causes a gradient of acceleration:[4]

$$\frac{dg}{dz} = - \frac{g_0}{L_0} \sin^3\theta \qquad (1)$$

where L_0 is the centrifuge arm length. We get, therefore:

$$\frac{d}{dz}(\rho g) = \frac{d\rho}{dz} \cdot g_0 - \rho_0 \left| \frac{dg}{dz} \right| \prec \frac{d\rho}{dz} \cdot g_0 \qquad (2)$$

So: (1) In the S case, $(d\rho)/(dz) < 0$ and the centrifugation overstabilizes the fluid. (2) In the D case, $(d\rho)/(dz) > 0$ and two competitive contributions appear, which could result in stabilizing the D case; if, in the above experiment on tin (dilatation coefficient $\alpha = 10^{-4}$ K^{-1}),

$$\cos^3\theta \succ \left(\frac{1}{\rho_0} \cdot \frac{d\rho}{dz} \right) \cdot L_0 \qquad (3)$$

is true, when the resultant acceleration is larger than a certain threshold, which depends on the axial temperature gradient as shown by Fig. 7.

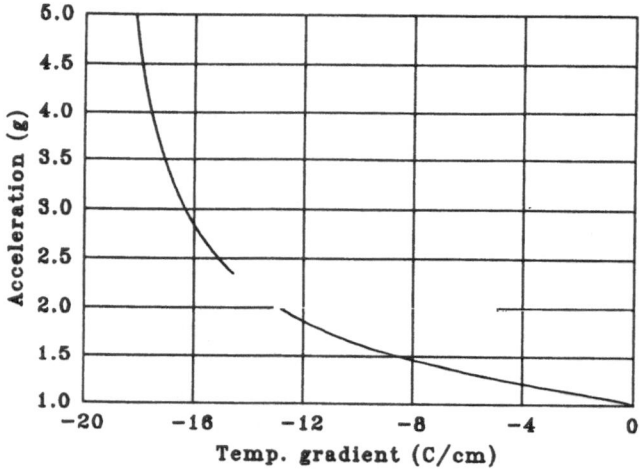

Figure 7. Acceleration required to annul the buoyancy effect of a given destabilizing temperature gradient in a tin melt on the Nantes centrifuge.

This simple model suggests the introduction of an appropriate Rayleigh number, taking into account the effective gravity field axial inhomogeneity, as well as the resulting sign of $(d/(dz))(\rho g)$:

$$Ra_{axial} = \frac{h^3}{vk}\left[\left(\frac{\rho}{\rho_0} \cdot g\right)_{top} - \left(\frac{\rho}{\rho_0} \cdot g\right)_{bottom}\right] \qquad (4)$$

where h is the height of the fluid, v and k are the momentum and thermal diffusivities, respectively. Note that Ra_{axial} decreases with increasing Ω (cf. the above observation where the axial temperature gradient decreases with increasing Ω). A negative value of this number would correspond to an effectively stabilized situation. This helps to understand why "nonaxisymmetric" centrifugation can provoke the reported reversible transitions.

CONCLUSIONS

The influence of centrifugation on thermal convection depends on the direction of the imposed temperature gradient, relative to the resultant acceleration vector. For the destabilizing configuration, we have shown that centrifugation suppresses the temperature fluctuations via a gradual reversible transition and forces an important reduction in melt temperature gradients. In the case of the stabilizing configuration, the melt temperatures are stable for all accelerations and the convection regime appears to be independent of centrifugation.

It was demonstrated that our nonaxisymmetric centrifugation procedure contains a stabilizing mechanism. The mechanism overstabilizes thermally stabilized fluids and is able to stabilize a thermally unstable fluid. This finding may help to suppress temperature fluctuations in crystal growth using a top seeding configuration. This is the case, in particular, for the well-known and industrially important Czochralski system.

Acknowledgement

This work was supported by the EEC Program BREU 0262 M under Project BE 3628–89, EEC Science Plan Bursary B/SCI–900366, DRET Project 89–153, and a subvention from the French Ministry of Research.

We are very grateful to J. Garnier and L.M. Cottineau of the Laboratoire Central des Ponts et Chaussées for their invaluable assistance and access to the Nantes centrifuge.

REFERENCES

1. "Proceedings of the First International Workshop on Material Processing in High Gravity," Dubna USSR, *J. Crystal Growth* 119 (1992).
2. W. Weber, G. Neumann, and G. Müller, *J. Crystal Growth* 100:145 (1990).
3. W.A. Arnold, W.R. Wilcox, F. Carlson, A. Chait, and L. Regel, *J. Crystal Growth* 119:24 (1992).
4. M.A. Fikri, G. Labrosse, and M. Betrouni, *J. Crystal Growth* 119:41 (1992).
5. H. Rodot, L.L. Regel, and A.M. Turtchaninov, *J. Crystal Growth* 104:280 (1990).
6. A. Chevy, *C.R. Acad. Sci. Serie II* 307:1147 (1988).
7. S. Chandrasekhar, "Hydrodynamic and Hydromagnetic Stability," Oxford University Press, London (1961).

GROWTH OF GaAs SINGLE CRYSTALS AT HIGH GRAVITY

Bojun Zhou[1], Funian Cao[1], Lanying Lin[1], Wenju Ma[2],
Yun Zheng[2], Feng Tao[2] and Minglun Xue[2]

[1]Institute of Semiconductors
Chinese Academy of Sciences
Beijing 100083, China

[2]Institute of Mechanics
Chinese Academy of Sciences
Beijing 100080, China

ABSTRACT

Two GaAs single crystals were grown on a centrifuge at 3g, 6g and 9g acceleration. Impurity striations and dislocations were observed for different gravity conditions. Temperature oscillations in molten Sn indicated that the conditions used to grow GaAs single crystals showed depressing temperature oscillations. A possible reason is given for the increase of dislocation density of GaAs with increasing centrifugal force in a sand-blasted quartz boat.

INTRODUCTION

Since the early work by Müller[1,2], many scientists believe that striations in semiconductor materials can be suppressed during growth of single crystals by using high gravity conditions. In addition, according to the results on space grown GaAs [3,4], we believe that gravity affects compositional homogeneity. We chose GaAs for our experiments because GaAs has become one of the most attractive electronic materials due to its excellent semi-insulating properties, which promise developments in high speed electronic and photonic devices.

The first GaAs single crystal was grown on a centrifuge in 1988 at 3g, 6g and 9g centrifugal acceleration. A Sn melt was used to simulate the temperature oscillations in a GaAs melt. The second GaAs single crystal was grown on a centrifuge in 1992.

EXPERIMENTS

All the GaAs single crystal growth and Sn simulation experiments were carried out on the same centrifuge. The centrifuge, 14 m in diameter, is owned by the Chinese Academy of Space Technology. A schematic diagram of the furnace is shown in Fig.1.

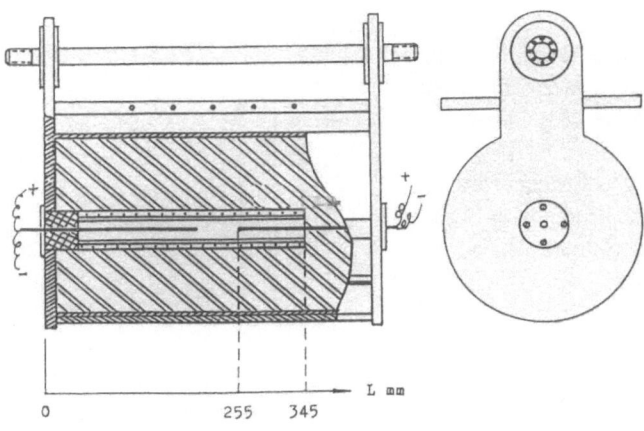

Figure.1. Schematic diagram of the furnace.
The right thermocouple is for temperature control, the left hermocouple is for measuring the temperature profile.

The furnace was hung on the end of one arm of the centrifuge by a rod at the top of the furnace. The axis of the furnace was perpendicular to the arm of the centrifuge, i.e. in the "horizontal" boat configuration, the furnace was rotated smoothly to keep the molten GaAs or Sn in the quartz boat. The furnace can sustain the strong wind during rapid rotation of the centrifuge, and can provide a suitable temperature profile for the growth of GaAs single crystals.

GaAs single crystals were grown by the gradient freeze technique. The electron concentration of the Te-doped GaAs single crystal was $(2-5) \times 10^{18}/cm^3$. There was additional As in the ampoule to supply an As pressure of one atmosphere during crystal growth. The sand-blasted quartz boat was 2 mm higher than a normal semicircular boat. Before sealing, the quartz tube with GaAs ingot and additional As was baked at 320°C for 30 minutes in a vacuum of 10^{-6} torr. The distribution of the temperature in the furnace and the location of the GaAs ingot in the furnace are shown in Figure 2. About half of the GaAs ingot was melted, and the furnace temperature was stabilized as the centrifuge was put into rotation. In order to control the freezing rate at 6-9 mm/hour, a cooling rate 0.2°C/min was used. Different centrifugal accelerations were used during the same GaAs single crystal growth: 3g for 1.5-2 hours, 6g for 1.5-2 hours and 9g for 1-1.5 hours. Molten Sn was used for measurement of temperature oscillations because temperature measurement in GaAs melt was difficult.

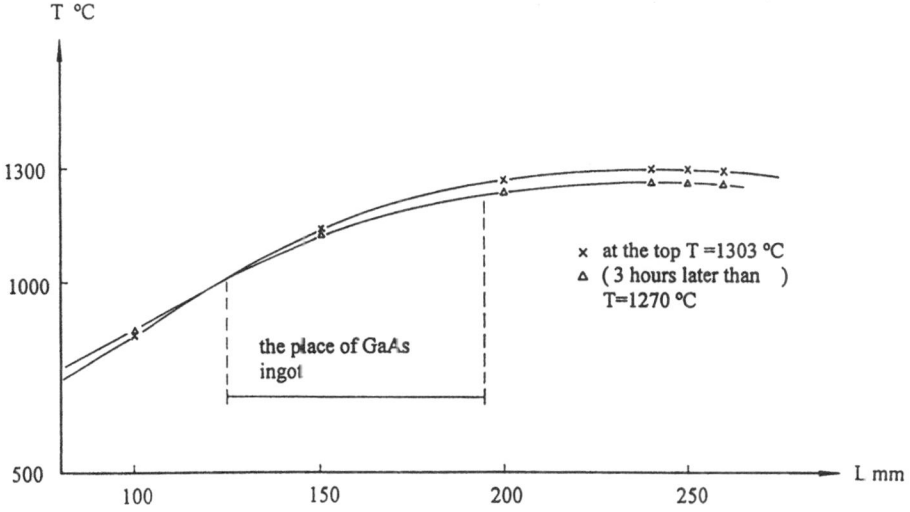

Figure.2 Temperature distribution in the furnace

Striations in GaAs were revealed by anodic etching of a {100} surface of a GaAs single crystal grown in high gravity. The dislocation density was revealed using the KOH method.

RESULTS AND DISCUSSION

Figure 3 shows two GaAs single crystals grown at 3, 6 and 9g in 1988 and in 1992. The roughness of the bottom surfaces increased with increasing centrifugal acceleration.

The surface of the sand-blasted quartz boat consisted of many small peaks, pits and small valleys. Under normal 1g, molten GaAs is supported on the peaks by surface tension. When the centrifugal acceleration was increased, the molten GaAs was pushed into the pits and valleys increasing the roughness of the bottom.

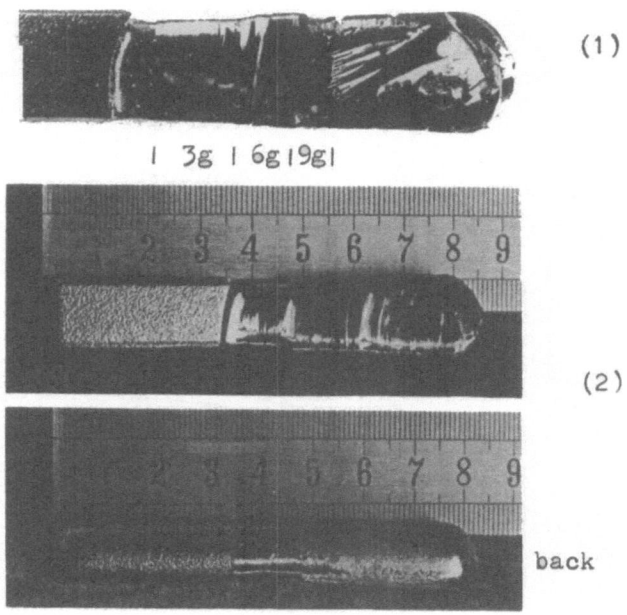

Figure.3. Photos of the first and second GaAs single crystals grown in centrifuge
Top: first. Middle and bottom: second.

During growth and cooling, the stress in the crystal should be larger in high gravity, because thermal expansion is different between quartz and GaAs. The experiment results agreed with this analysis. The dislocation densities for different gravities are shown in Fig.4. Cathodoluminescence (CL) topography [5] of the first GaAs single grown on the centriguge gave the same dislocation density. According to this explanation we cannot say that the quality of GaAs single crystal grown in high gravity would be worse if we didn't use a sand-blasted quartz boat. It is suggested that the boat material should be changed and that the surface of the boat should be smooth. We made one experiment in high gravity using a graphite boat but didn't obtain a GaAs single crystal.

The striations in Fig.5(1) are for 1g and are the most dense. This indicates that the homogeneity of GaAs was improved by growth at high gravity. Comparing Fig.5.(3) to

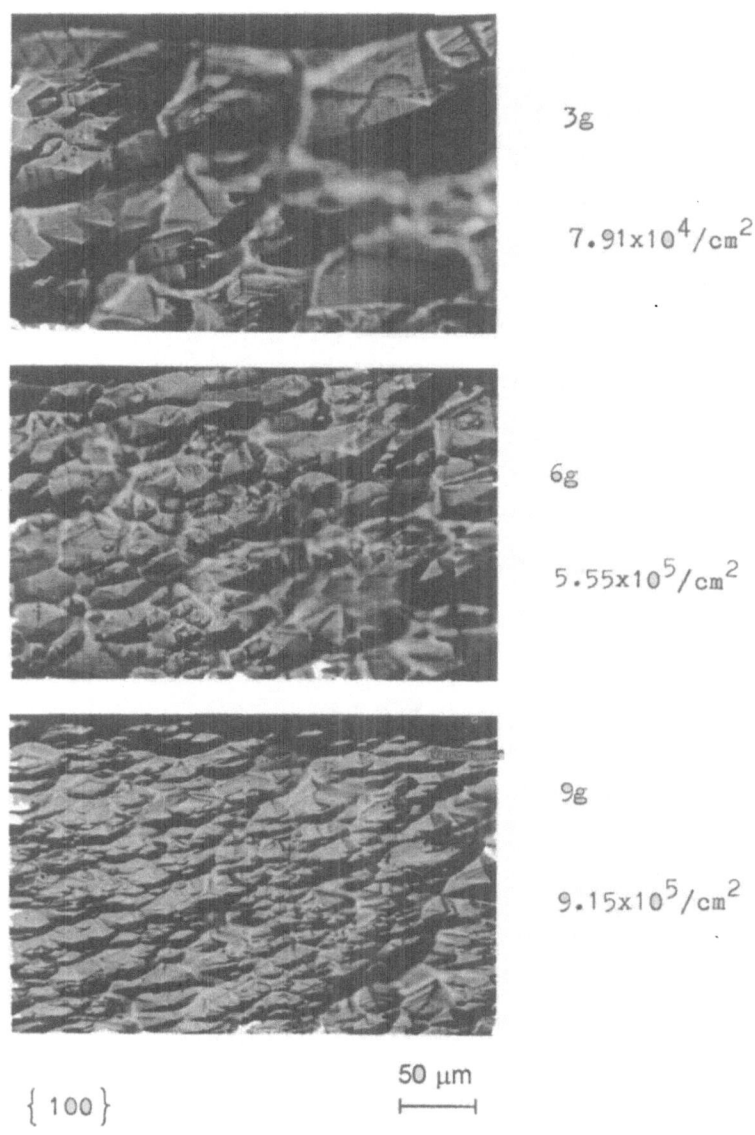

3g

$7.91 \times 10^4/\text{cm}^2$

6g

$5.55 \times 10^5/\text{cm}^2$

9g

$9.15 \times 10^5/\text{cm}^2$

{ 100 }

50 µm

Figure.4. Dislocation etch pits in the second GaAs single crystal
grown on the centrifuge

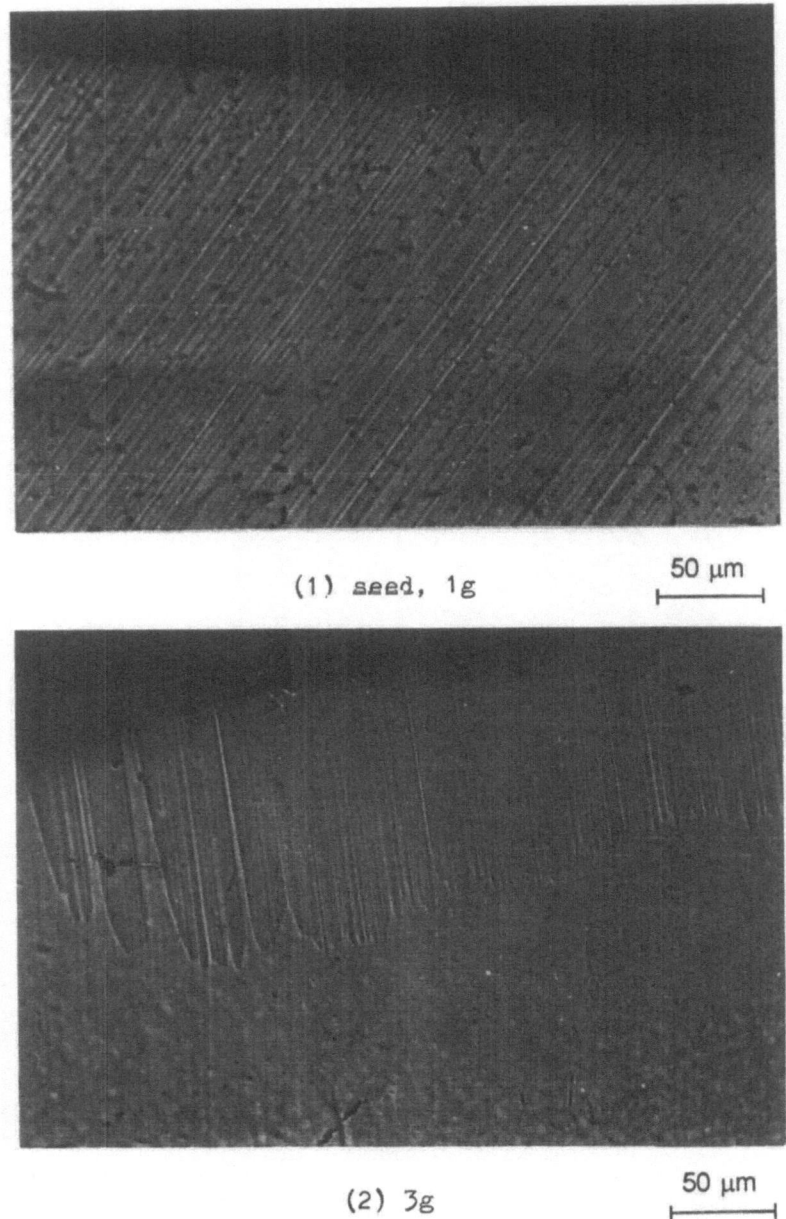

(1) seed, 1g

50 μm

(2) 3g

50 μm

Figure.5. Striations in second GaAs single crystal grown in centrifuge
(1) seed,1g; (2) 3g; (3) 6g; (4) 9g.

(3) 6g   50 μm

(4) 9g   50 μm

Fig.5.(4), it is seen that the striations at 9g are less dense than at 6g. The striations in the upper part of Fig.5.(2) are in a faceted region.

The molten Sn simulation results showed that: 1. The temperature gradient decreased with increasing rotation velocity. 2. When the higher temperature part of the ampoule was set in front (against the wind direction) during rotation of the centrifuge ($\omega>0$), the flow in the melt was stable, when $\omega<0$, the flow was unstable. The temperature fluctuations in molten Sn versus rotation sense of the centrifuge are shown in Fig.6. Note that our two GaAs single crystals grown on the centrifuge were grown with $\omega>0$.

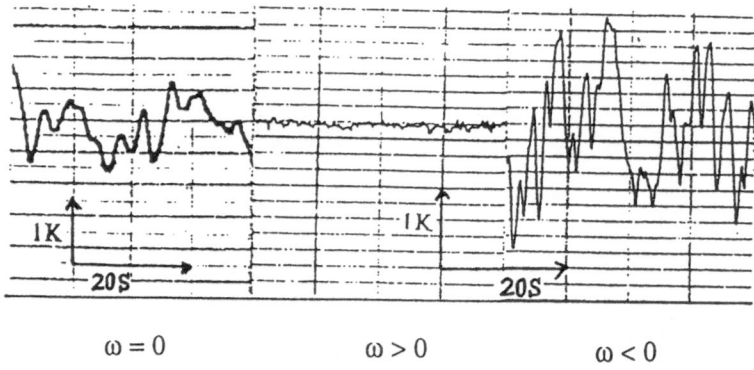

$$\omega = 0 \qquad \omega > 0 \qquad \omega < 0$$

Figure.6. Temperature fluctuations in molten Sn with different rotation senses of the centrifuge. $G_t=12K/cm$.

ACKNOWLEDGEMENTS

This work was supported by the Chinese Academy of Sciences. The centrifuge facility was provided by the Chinese Academy of Space Technology.

REFERENCES

1. G. Müller, E.Schmidt and P.Kyr, *J.Crystal Growth,* 49 (1980) 387-395.
2. G. Müller and G.Neumann, *J.Crystal Growth,* 59 (1982) 548-556.
3. B. J. Zhou, X.R. Zhong, F.N. Cao, L.Y. Lin, D.A. Da, K.L. Wu, L.F. Huang, S.H. Zheng and X. Xie, *Chinese Journal of Semiconductors,* 9 3 (1989) 309-315
4. L.Y. Lin Group and D.A. Da Group, *Mater. Sci. Forum,* 50 (1989) 183.
5. X.R. Zhong, B.J. Zhou, O.M. Yan, F.N. Cao, C.J. Li and L.Y. Lin, *J. Crystal Growth,* 119 (1992) 74-78.

RESPONSE OF TEMPERATURE OSCILLATIONS
IN A TIN MELT TO CENTRIFUGAL EFFECTS

W.J. Ma,[1] F. Tao,[1] Y. Zheng,[1] M.L. Xue,[1] B.J. Zhou[2] and L.Y. Lin[2]

[1]Institute of Mechanics
 Chinese Academy of Sciences
 Beijing 100080, China
[2]Institute of Semiconductors
 Chinese Academy of Sciences
 Beijing 100083, China

ABSTRACT

An experimental study was conducted on the flow and temperature oscillations of molten tin in a horizontal boat under centrifugation. A longitudinal temperature gradient was applied to an open boat containing molten tin so that convection was generated with a known direction of the basic flow. The experimental results demonstrated that both effects of centrifugal and Coriolis forces exist due to the rotation of the centrifuge. The former enhances the convection, decreasing the temperature gradient in the melt. The latter shows different influences on the flow stability depending on the rotation sense of the centrifuge. Temperature fluctuations in the melt are considerably retarded, provided the centrifuge rotation is in the same sense as the convection roll in the melt.

INTRODUCTION

Melt growth of crystals is widely used in industry. During the crystal growth process, convection exists within the melt. Convection may be beneficial for transport of heat and mass, but unsteady convection causes temperature fluctuations in the melt. These, in turn, give rise to a fluctuating concentration of solute in the crystal, forming so-called impurity striations. In order to avoid such convection-induced microinhomogeneities, various measures have been taken to maintain a steady state of buoyancy convection or suppress unsteady convection, such as changes of the melt dimension and geometry, reduction of the temperature gradient, application of static magnetic fields, and reduction of gravity (space experiments). High gravity produced by a centrifuge has been used to grow crystal in recent years. Suppression of unsteady convection and doping striations with centrifugal acceleration was reported for upside-down vertical Bridgman growth and for horizontal zone melting on a centrifuge.[1] We have grown GaAs single crystals by a horizontal gradient freeze method on a large centrifuge. Our preliminary results demonstrated that impurity striations in GaAs grown under centrifugal acceleration became weak and indistinct.[2] In order to make clear the effects of centrifugation on crystal growth, an experiment was conducted to simulate temperature oscillations in the GaAs melt. In this paper we address the preliminary experimental results for temperature fluctuations of liquid tin, whose Pr number is similar to that of molten GaAs. The liquid tin was held in a horizontal boat mounted on a centrifuge. Both centrifugal and Coriolis forces had significant influences on the melt flow.

EXPERIMENTAL ARRANGEMENT

The experiment was carried out on the same centrifuge as described earlier.[2] It has an arm with a length of 7 m and an available centrifugal acceleration up to 30 times earth's gravity **g**. A horizontal cylindrical furnace was hung on the end of the arm via a bearing with its axis perpendicular to both the centrifuge arm and earth's gravity (see Fig. 1 of reference 2). A sealed quartz tube containing an open boat was located in a resistance heater, which established a longitudinal temperature gradient. A tin melt 80 mm long and 8 mm high was placed in the boat. A temperature gradient was applied so that thermal convection was generated with the flow ascending at the end with higher temperature and descending at the end with lower temperature. Thermocouples with protectively coated beads were dipped into the melt and fixed at the positions shown in Fig. 1. The temperature gradient and temperature fluctuations in the tin melt were measured by thermocouples A, B and C (NiSi-NiCr, diameter 1 mm). The temperature gradient G is defined by the temperature difference between two thermocouples A and B divided by the distance between them. The temperature in the furnace was monitored by the thermocouple M and controlled by a control unit during the whole experiment. The temperature profile in the melt was determined by the furnace temperature and influenced by the rotation of the centrifuge. The power and the amplified thermocouple signals were transmitted via slip rings.

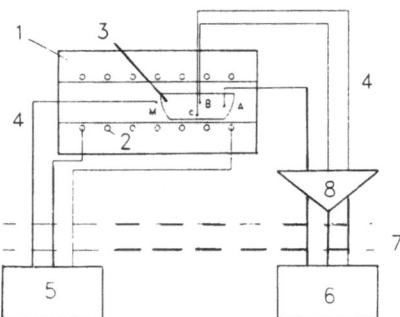

Figure 1. Sketch of the experimental set-up. 1 furnace, 2 resistance heater, 3 boat, 4 thermocouple, 5 thermal control unit, 6 A/D & chart recorder, 7 slip rings, 8 amplifier.

RESULTS AND DISCUSSION

Before running the centrifuge the furnace was heated and the temperature gradient in the furnace increased. The temperature fluctuations in the melt were pronounced, as shown in Fig. 2, which is a plot of the temperature measured at point C in Fig. 1 for different temperature gradients in the melt. This is consistent with the well-known argument that an increase of the longitudinal temperature gradient will produce temperature fluctuations with increasing amplitude and frequency.[3]

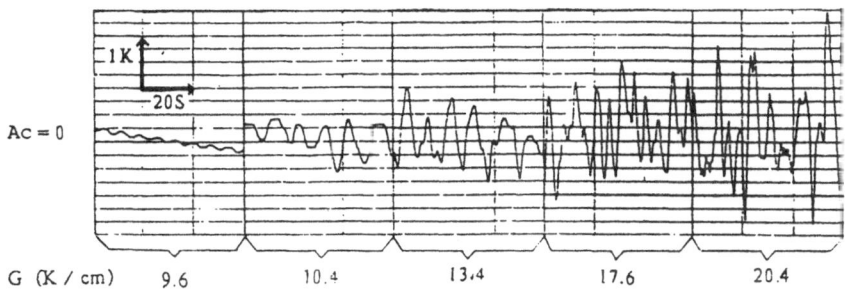

Figure 2. Temperature records for various temperature gradients G in the melt, without centrifuge rotation. G is calculated based on the measured temperature difference between thermocouples A and B.

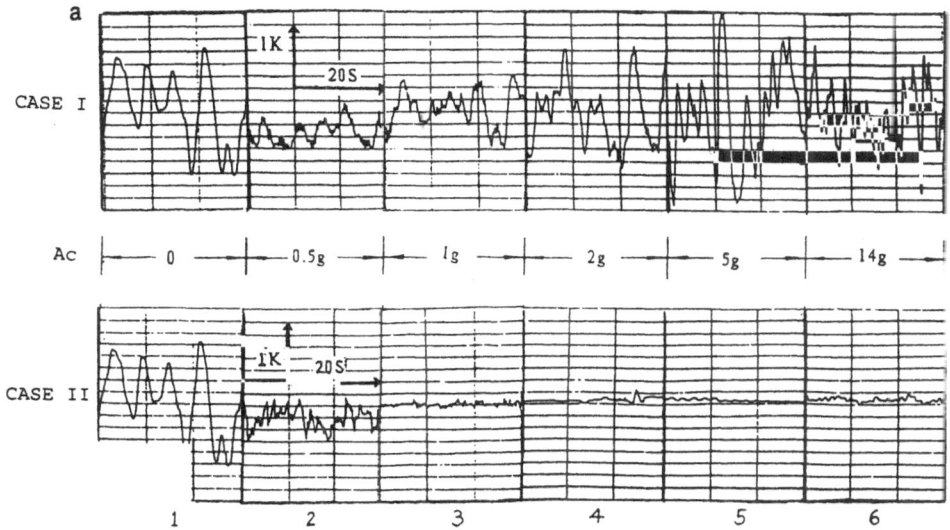

Figure 3. Records of temperature fluctuations (a), and the power spectra (b), for two different rotation senses of the centrifuge. A_c is centrifugal acceleration, g is earth's gravity. (continued next page)

When the temperature gradient in the melt reached about 16 K/cm, the centrifuge was turned on and the rotation rate was raised stepwise, with each step stabilized for about ten minutes. The axial temperature gradient along the furnace was kept constant. Two series of tests were performed with respect to the rotation sense of the centrifuge. Figure 3 is a plot of temperature fluctuations measured by thermocouple C for different centrifugal accelerations and their power spectra for two different rotation senses of the centrifuge. It is seen from Fig. 3(a) that for case I (the direction of the convective roll in the melt is opposite to that of the centrifuge rotation), the amplitude of temperature oscillations increases with increasing centrifugal acceleration A_c, and the oscillations of higher frequencies also increase Fig. 3(b). For case II (both directions of the melt convective roll and the centrifuge rotation are in the same sense), the temperature oscillations are dramatically suppressed when $A_c > 1g$.

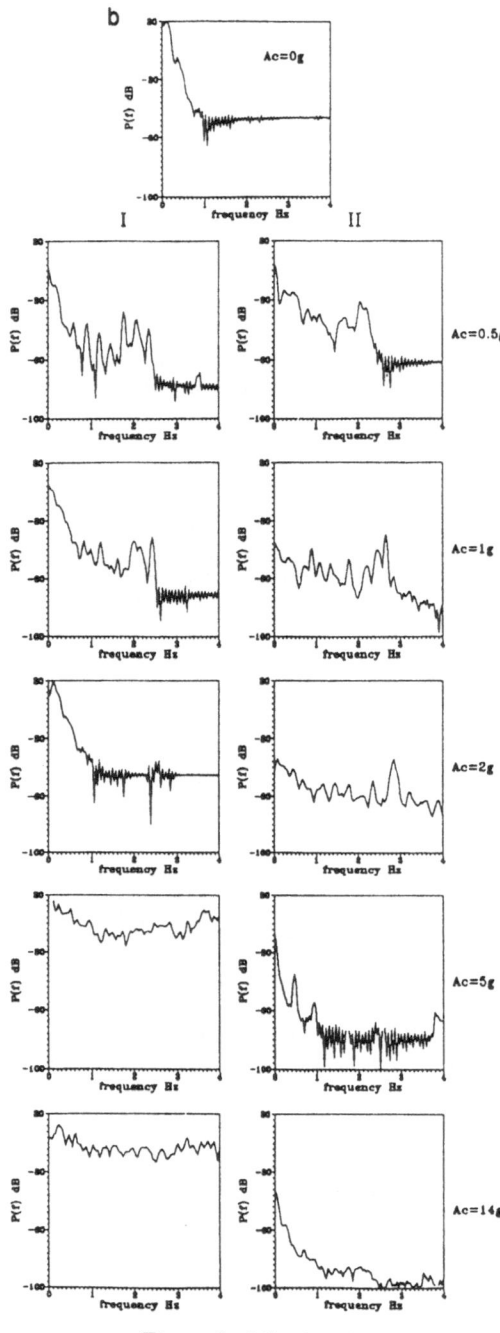

Figure 3. (Continued)

Figure 4 shows the measured temperature gradient G in the melt versus the centrifugal acceleration A_c. In both cases the temperature gradient decreased with increasing centrifugal acceleration. This means that the centrifugal force acts as an increased gravity, enhancing the buoyancy-driven convection and heat transfer, and decreasing the temperature gradient in the melt. It is also seen that for case II the temperature gradients decrease much faster than for case I. The reason will be discussed later.

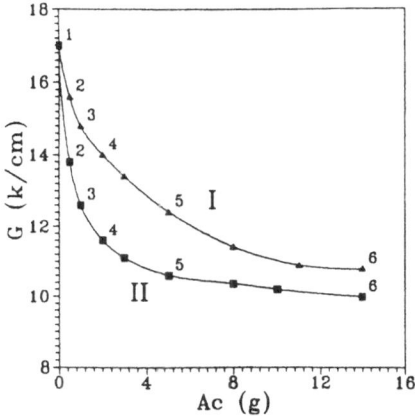

Figure 4. Measured temperature gradient G in the melt versus centrifugal acceleration A_c for two cases. The numbered points correspond to those labeled in Fig. 3.

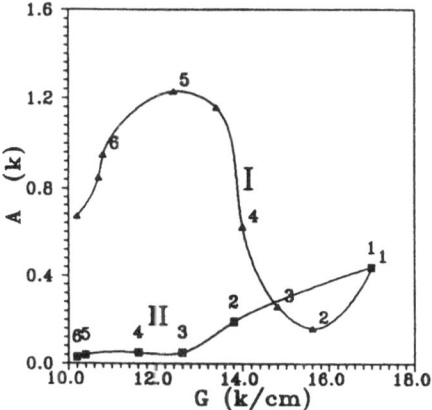

Figure 5. Measured average amplitude A of temperature fluctuations versus temperature gradient G in the melt for two cases. The numbered points correspond to those labeled in Fig. 3.

In Fig. 5 we plot the average amplitude of temperature fluctuations versus the temperature gradient for two different rotation senses. The temperature fluctuations decreased in the initial stage of rotation of the centrifuge for both cases I and II (point 1 to 2 in Fig. 5). The fluctuations continued to decrease with decreasing temperature gradient for case II, but increased for case I after point 2. The dependence of the experimental results on the rotation sense of the centrifuge may be explained by the influence of the Coriolis force as stated by Müller *et al.*[1,4] For case II the component of the angular velocity of the centrifuge perpendicular to the resultant acceleration vector produces a component of Coriolis force that acts outwards with respect to the convection roll, forcing the fluid to stream along elongated paths. As a result, heat transport is increased and temperature oscillations are damped. For case I, however, the Coriolis force pushes the fluid towards the center of the convection roll causing the flow to become more unstable. By comparing Fig. 5 with Fig. 2, one can find that even for the same temperature gradient G, the amplitude of the oscillations for case II are still smaller than without rotation.

Unlike the vertical Bridgman geometry with top seeding, the horizontal Bridgman configuration has a pre-known direction of the main convection roll subject to the applied temperature gradient. Therefore no bifurcation phenomenon[4] occurs. This allows one to suppress unwanted fluctuations of temperature during crystal growth by setting the orientation of the furnace or selecting the rotation sense of the centrifuge as required for case II. It should be mentioned that in the horizontal zone melting configuration there are two opposite convective rolls. Both cases I and II may exist simultaneously. Thus a proper orientation of the melt zone (i.e. the zone part adjacent to the growing crystal front is in the situation of case II) seems necessary.

CONCLUDING REMARKS

The experimental results demonstrate that there are two effects that occur due to the rotation of the centrifuge. One is the effect of the centrifugal force, which enhances the convection and decreases the temperature gradient. Another is the effect of Coriolis force, which leads to different influences on the temperature oscillations depending on the rotation senses of the centrifuge and the convection roll in the melt. This fact has implications for improving crystal growth. A further experiment with fixed thermal conditions at the boat ends and a larger acceleration range is planned.

Acknowledgments

This work was supported by the National Natural Sciences Foundation and the special financial aid provided by the Chinese Academy of Science. The authors acknowledge BISEE for offering the use of the centrifuge facility.

References

1. G.Müller, G.Neumann and W.Weber, *J. Crystal Growth* 119:8 (1992).
2. X.R.Zhong, B.J.Zhou, Q.M.Yan, F.N.Cao, C.J.Li, L.Y.Lin and W.J.Ma, Y.Zheng, F.Tao,M.L.Xue, *J. Crystal Growth* 119:74 (1992).
3. G.S.Cole, W.C.Winegard, *J. Inst. Metals* 93:153 (1964).
4. W.Weber, G.Neumann and G.Müller, *J. Crystal Growth* 100:145 (1990).

UNSTEADY THERMAL CONVECTION OF MELTS IN A 2-D HORIZONTAL BOAT IN A CENTRIFUGAL FIELD WITH CONSIDERATION OF THE CORIOLIS EFFECT

F. Tao,[1] Y. Zheng,[1] W.J. Ma[1] and M.L. Xue[1]

[1] Institute of Mechanics
Chinese Academy of Sciences
Beijing 100080, China

ABSTRACT

A rotating centrifuge introduces the centrifugal acceleration and the Coriolis force acting on melts while melt growth is being carried out in the centrifuge. These two forces influence melt convection and, in turn, modify the transport of dopant and impurities. In this paper the effects of varying the centrifugal acceleration and the Coriolis force were studied numerically. We paid attention to unsteady thermal convection of melts in a two-dimensional rectangular boat with relevance to crystal growth in a centrifuge by horizontal Bridgman technique. The mathematical model was constructed by the continuity, Navier-Stokes and energy equations with the Boussinesq approximation, which was solved by the finite control volume method with fully implicit, steady, time-marching, central-difference discretization. The calculations based on the simplified model reveal that the centrifugal acceleration enhances buoyancy force, which may dominate the convection and induce oscillation, and the Coriolis force may stabilize or destabilize the flow depending on the rotation sense of the centrifuge. This numerical results as well as the experiments of temperature measurement[15] give a satisfactory explanation of the results described previously[12,13].

INTRODUCTION

In crystal growth from melts, convection in the melt is of great importance in determining the quality and compositional uniformity of the grown crystal because it is usually oscillatory or fluctuating flow that causes compositional striations in the crystal. In order to grow single crystal within the environment of a stabilized melt crystal grower has to treat the growth system carefully. However, one only with experience obviously can not satisfy the increasing demand for more perfect single crystals (e.g. Si and GaAs). Thus,

Materials Processing in High Gravity, Edited by L.L. Regel
and W.R. Wilcox, Plenum Press, New York, 1994

scientists, such as Carruthers[1], Langlois[2] and Müller[3], exerted themselves much to study the role of melt convection and the relevant mechanisms in crystal growth.

Gravitational effect is in general important, and even dominant, to thermal convection of melts unless melt growth is carried out in microgravity environment. It has been well known that growth of crystals in such environment has more advantages than on earth because microgravity effect hinders buoyancy convection. Unexpectedly, in recent years, when Müller and his colleagues studied the gravitational effect on crystal growth, they found that single crystal also could be grown under high gravity, i.e. the centrifugal acceleration produced by a centrifuge.[3-9] Using the centrifuge in their laboratory, they grew InSb and GaSb crystals free of doping striations by an upside-down, tilted Bridgman method and a horizontal zone melting technique. Their finding now fascinates many crystal growers and scientists who have begun to do research work on this interesting subject.[10-15]

From the initial purpose of their experimental research it is natural for Müller et al. to attribute the results to the effect of high gravity. However, it must be noted that a centrifugal field does not produce high gravity monotonically. Besides the centrifugal acceleration, the Coriolis force simultaneously occurs in the melt and alters convection in the same centrifugal field. Certainly, when these two forces act in concert the flow structures are not in a simple form but quite complex. Thus, a more careful investigation on this problem is still needed. After a series of research work through experiments and numerical simulations, Müller et al. observed that steady convection could occur in the melt, and concluded in their review paper[9] that the steady phenomena were not due to high gravity but the effect of the Coriolis force. For a different arrangement Regel et al. presented a dissimilar opinion on the subject.[10,11,14]

Unlike the methods used by Müller et al. and Regel et al., we take special interest in the horizontal Bridgman technique. Results of the GaAs crystals grown in a large centrifuge showed that the impurity striations became weak or less dense than their original ones[12,13]. In order to obtain more understanding of convection under high gravity we carried out a model experiment of temperature measurement with molten Sn in the same centrifuge.[15] Interesting phenomena due to the Coriolis effect were also observed in this model experiment.

However, up to now we have only limited experimental data, which led to an incomplete understanding of the behavior of the melt in the horizontal Bridgman technique in the centrifuge. To understand and explain the melt behavior comprehensively, we need more information on the flow field. For the horizontal Bridgman technique, there is a problem of finding a method for direct observation of the melt flow, not to mention the difficulties in performing experiments in the centrifuge. Thus, we carried out a numerical simulation to serve these requirements.

In this research the effects of varying the centrifugal acceleration and the Coriolis force were studied numerically. We paid attention to time-dependent thermal convection of melts in a two-dimensional, rectangular configuration of open boat with relevance to crystal growth in a centrifuge by horizontal Bridgman technique. In section 2 we describe the physical problem and discuss construction of the numerical model. Next, in section 3 we introduce time-dependent continuity, Navier-Stokes (with Boussinesq approximation) and energy equations in primitive variables. After section 3 we present the grid system and the algorithm for the calculation. Section 5 gives the numerical results. Finally, we draw some conclusions on the phenomena of convective flow in horizontal Bridgman technique in a centrifugal field.

DESCRIPTION OF THE PHYSICAL MODEL

The experimental arrangement of horizontal gradient freeze method in the large centrifuge described previously[12,13] is schematically pictured in Figure 1. An open boat in the shape of a semi-circular horizontal cylinder was positioned within a horizontal furnace and subjected to a temperature gradient. To prevent the melt from spilling out of the boat during experiments the furnace, mounted at the end of the arm of the centrifuge, was hinged on the beam so that it would align itself with the resultant gravity vector **g**. Therefore, the upper surface of the melt was flat and stationary with respect to **g**.

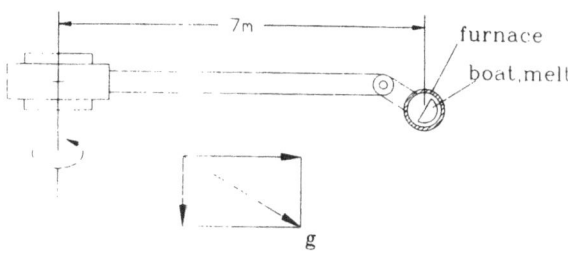

Figure 1. Sketch of crystal growth by the horizontal Bridgman technique in the large centrifuge.

For melt convection in a horizontal boat, a two-dimensional simulation is usually considered necessary to show some of the main characteristics of the real flow structure. However, at present, it is difficult for us to carry out this calculation because the buoyancy effects of both earth gravity and centrifugal acceleration, along with the Coriolis effect, may result in three-dimensional convection.

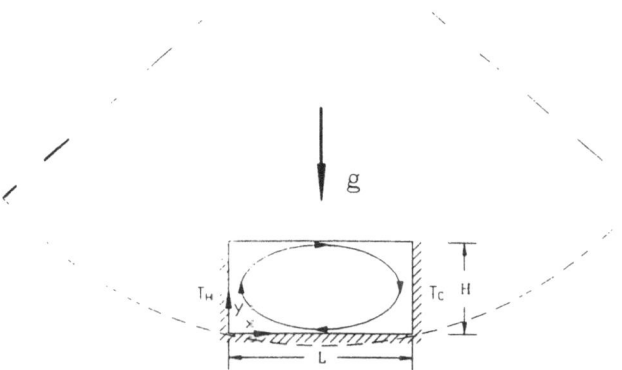

Figure 2. Configuration of the two-dimensional horizontal rectangular boat for the simulation of melt flow in a centrifugal field.

We now set some assumptions for the two-dimensional model. First we eliminate earth gravity and assume that the buoyancy effect on melt convection is solely caused by centrifugal acceleration. If, for example, the centrifugal acceleration is six times earth gravity, the furnace will lie on a position near the plane of the arm of the centrifuge with the angle less than 10°. In this case, the buoyancy should be regarded as dominated by the centrifugal acceleration. Under the above assumption, the influence of earth gravity therefore could be neglected and the centrifugal acceleration would be perpendicular to the melt surface. This two-dimensional idealized model is shown in Figure 2. It has a rectangular configuration of length L and thickness H. The upper boundary is the free surface of the melt. Both vertical side-walls are assumed to be held at different temperatures, T_H at the left and T_C at the right, respectively, with $T_H > T_C$. Here, we do not consider the case where T_C is lower than the melting-point T_m, but rather $T_C > T_m$. Then, standing on the centrifuge one would observe that the melt moves in a clockwise sense for either direction of centrifuge rotation.

The Coriolis force entailed in a centrifugal field as well as the buoyancy effect of the centrifugal acceleration has also a vital influence on the convection. Indeed, if the Coriolis force did not occur in the centrifugal field the simulation could be treated as those described by Crochet et al.[16] and Fontaine et al.[17] for conventional growth. Analysis shows that its effect cannot be ignored. From the definition of the Coriolis force, $f_c = -2\omega \times u$, two different effects of the Coriolis force can act on melt flow. For example, if only one single convective cell occurs in the boat as in Figure 2, these effects, I and II, will take place as shown in Figure 3. They depend strongly on the rotation sense of the centrifuge, similar to that described by Müller et al.[9]. We discuss the results related to this problem in section 5.

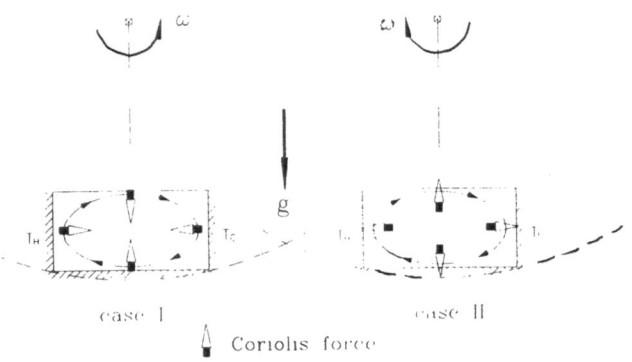

Figure 3. The two patterns of the Coriolis effect on melt convection in the horizontal Bridgman boat, depending on the direction of rotation of the centrifuge.

GOVERNING EQUATIONS AND BOUNDARY CONDITIONS

Let us consider the simplified model in Figure 2. The boat and the two-dimensional coordinate system are both fixed on the arm of the centrifuge, and turn around the axis of the centrifuge when the centrifuge operates. The fundamental flow field variables are the two velocity components u and v, pressure p, and temperature T. The governing equations are:

$$\nabla \cdot \mathbf{u} = 0, \tag{1}$$

$$\frac{\partial \mathbf{u}}{\partial t} + (\mathbf{u} \cdot \nabla)\mathbf{u} = -\frac{1}{\rho}\nabla p + \nu\nabla^2\mathbf{u} + \beta\mathbf{g}(T - T_C) - 2\omega \times \mathbf{u}, \tag{2}$$

$$\frac{\partial T}{\partial t} + (\mathbf{u} \cdot \nabla)T = \kappa\nabla^2 T, \tag{3}$$

where **u** denotes the velocity vector of the flow field, ω the angular velocity vector of the centrifuge, ρ the density of the melt, ν the kinematic viscosity, β the volumetic expansion coefficient and κ the thermal diffusivity. Equations 1-3 are the continuity equation, Navier-Stokes equation, and energy equation, respectively. In equation 2 the validity of the Boussinesq approximation is presumed. The term $-2\boldsymbol{\omega} \times \mathbf{u}$ is the Coriolis force.

In order to obtain a general understanding of melt behavior, we perform the simulation in a dimensionless form. However, it must be noted that there is no obviously typical velocity scale for us to select. Necessarily it should be obtained from momentum balance. Since the buoyancy force dominates convection in the present problem, the magnitudes of terms $(\mathbf{u} \cdot \nabla)\mathbf{u}$ and $\beta\mathbf{g}(T - T_C)$ in equation 2 should be considered the same. Then we have:

$$(\mathbf{u} \cdot \nabla)\mathbf{u} \sim \beta\mathbf{g}(T - T_C), \tag{4}$$

or in terms of typical parameters equation 4 become:

$$\frac{UH}{\nu} = O(Gr^{1/2}), \tag{5}$$

where H is the typical length scale, U the typical velocity scale, and Gr the Grashof number defined as:

$$Gr = \frac{\beta\mathbf{g}(T_H - T_C)H^3}{\nu^2}. \tag{6}$$

Thus, the dimensionless variables are the following:

$$\mathbf{u}^* = \left[\frac{Gr^{-1/2}H}{\nu}\right]\mathbf{u}, \qquad x^* = \left[\frac{1}{H}\right]x, \qquad t^* = \left[\frac{Gr^{1/2}\nu}{H^2}\right]t,$$

$$p^* = \left[\frac{Gr^{-1}H^2}{\rho\nu^2}\right]p, \qquad T^* = \frac{T - T_C}{T_H - T_C}, \tag{7}$$

where the variables with the asterisk * are dimensionless. For convenience we drop * so that the dimensionless equations become:

$$\frac{\partial u}{\partial x} + \frac{\partial v}{\partial y} = 0, \tag{8}$$

$$\frac{\partial u}{\partial t} + u\frac{\partial u}{\partial x} + v\frac{\partial u}{\partial y} = -\frac{\partial p}{\partial x} + Gr^{-1/2}\nabla^2 u + v\frac{|Ta|^{1/2}}{Gr^{1/2}}\text{sign}(\omega), \tag{9}$$

$$\frac{\partial v}{\partial t} + u\frac{\partial v}{\partial x} + v\frac{\partial v}{\partial y} = -\frac{\partial p}{\partial y} + Gr^{-1/2}\nabla^2 v + T - u\frac{|Ta|^{1/2}}{Gr^{1/2}}\text{sign}(\omega), \tag{10}$$

$$\frac{\partial T}{\partial t} + u\frac{\partial T}{\partial x} + v\frac{\partial T}{\partial y} = \frac{1}{PrGr^{1/2}}\nabla^2 T, \tag{11}$$

$$\nabla^2 = \frac{\partial^2}{\partial^2 x} + \frac{\partial^2}{\partial^2 y}. \tag{12}$$

The parameters appearing in equations 9-11 are the Prandtl number,

$$Pr = \frac{\nu}{\kappa}, \tag{13}$$

and the Taylor number,

$$Ta = \frac{4\omega^2 H^4}{\nu^2} \, \text{sign}(\omega). \tag{14}$$

Equation 14 is a specialized definition for the Taylor number used in this study. As described in section 2 two different patterns of the Coriolis effect occur when the rotation direction of the centrifuge is changed. If Ta takes its normal definition by removing sign(ω) from equation 14 the Coriolis effect will be expressed unclearly. From equation 14, according to the definition of ω in the present coordinate system, ω is a positive and Ta>0 when the centrifuge rotates counter-clockwise; whereas Ta<0 when the centrifuge runs in a clockwise direction.

Generally, the kinetic and thermal boundary conditions for crystal growth are difficult to propose adequately. In this work we adopt those simplified boundary conditions considered by Crochet et al.[16]. On all the walls of the boat the no-slip condition is applied, except on the free-surface where a stress-free condition is used. The upper and lower boundaries are subjected to a linear temperature gradient. At both ends of the boat the temperature is held constant. Therefore, the boundary conditions in dimensionless form are:

$$x = 0, 0 \le y \le H: \qquad u = 0, v = 0, T = 1; \tag{15}$$

$$x = L, 0 \le y \le H: \qquad u = 0, v = 0, T = 0; \tag{16}$$

$$0 \le x \le L, y = 0: \qquad u = 0, v = 0, T = 1 - x/L; \tag{17}$$

$$0 \le x \le L, y = H: \qquad \frac{\partial u}{\partial y} = 0, v = 0, T = 1 - x/L. \tag{18}$$

NUMERICAL METHOD

The SIMPLE algorithm was employed for the simulation. We use the finite volume method with the pressure-velocity correction technique to solve the governing equations 8-11. The finite volume method divides the flow regimes into small control volumes, also called a staggered grid, illustrated in Figure 4. The pressure p and the temperature T are defined at the center, and the velocity components u, v on the faces of each control volume. The pressure-velocity correction, which is used to improve a guessed pressure distribution and to solve the velocity field, is derived via the continuity equation. A detailed description can be seen in reference[18].

In the process of formulating the discretization equations from the governing equations one problem encountered is how to discretize the convective and diffusive term exactly. Four proposed schemes are[18]: the central difference scheme, the upwind scheme, the hybrid scheme and the power-law scheme. The easy way to judge which scheme is best is to compare the computed results with those obtained by other numerical methods. For the present work we chose the central difference scheme. A comparison of these four schemes is presented in the next section.

The employed numerical method has the following characteristics: staggered control volumes, central-difference discretization for diffusive and advective fluxes, and fully implicit steady and time-marching schemes.

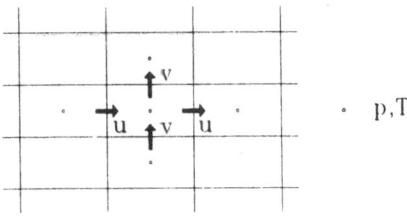

Figure 4. The staggered finite control volume and locations of variables u, v, p and T.

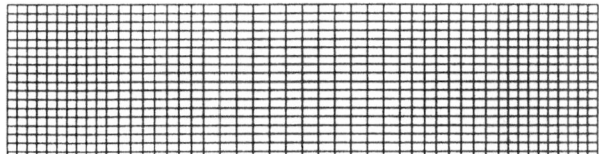

Figure 5. The 50×20 nonuniform mesh used in this study.

RESULTS AND DISCUSSION

In this section we consider the case of a constant Prandtl number, Pr=0.02, and an aspect ratio L/H=4. A 50×20 nonuniform mesh was designed to handle the high gradients near the vertical walls of the boat, as shown in Figure 5. The time-dependent convection is started from rest and the melting-point temperature is assumed everywhere at t=0. The time step for the time-marching scheme is 0.02.

The flow and temperature fields of the melt are displayed by using streamlines and isotherms. The stream function ψ is defined via the relations:

$$\frac{\partial \psi}{\partial y} = u, \qquad \frac{\partial \psi}{\partial x} = -v. \tag{19}$$

The time-dependent behavior of the melt can be found from the kinetic energy:

$$K = \frac{1}{2} \int_{\Omega} (u^2 + v^2) d\Omega \tag{20}$$

where Ω is the melt flow domain.

First of all, we excluded the Coriolis force from the problem and studied only the effect of high gravity. Without the Coriolis effect (i.e. Ta=0) our model is much similar to those described in references[16,17]. Therefore, we can easily make a comparison between the algorithm of this study and the finite element method[16] or the Tau-Chebyshev pseudospectral method[17]. This comparison helped us to catch on more to the four schemes described in the above section. The results at Ta=0 computed by the four schemes are shown in Figure 6.

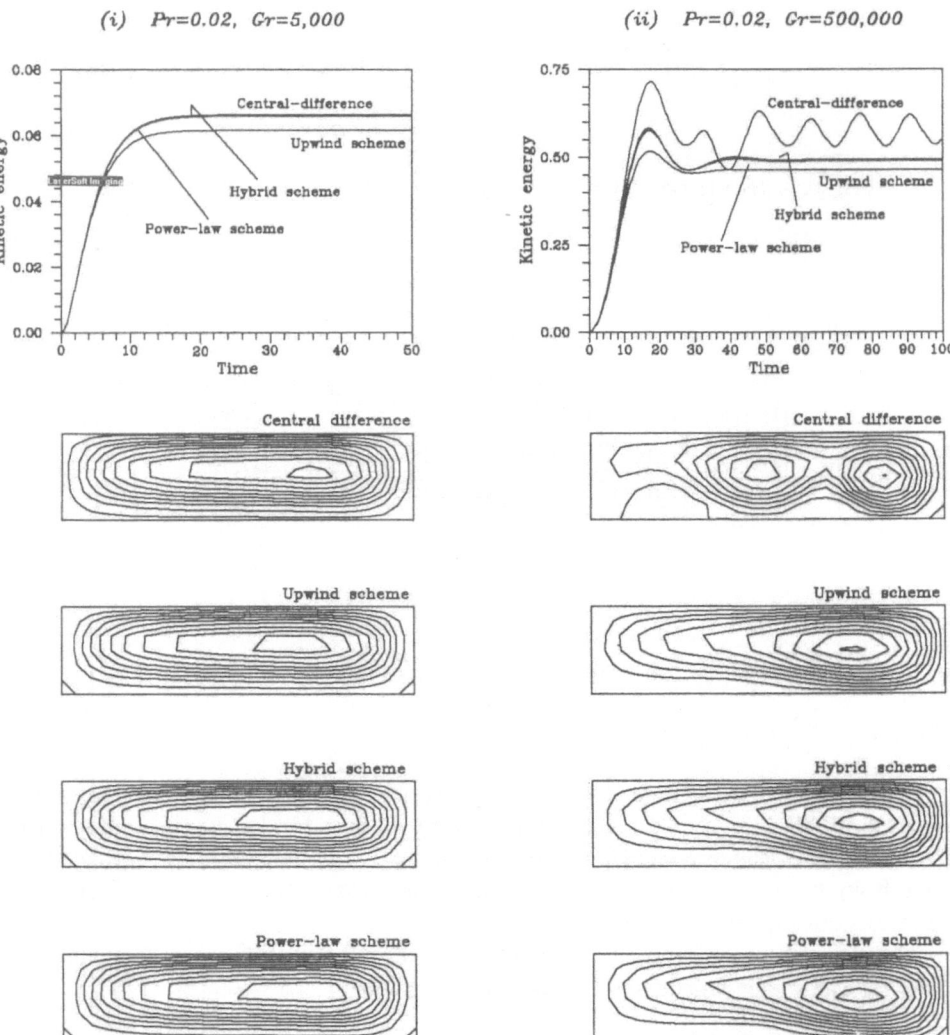

Figure 6. A comparison among four schemes: the central difference scheme, the upwind scheme, the hybrid scheme, and the power-law scheme. $Gr=5 \times 10^3$ and 5×10^5 at Ta=0 and Pr=0.02. The streamlines are plotted at t=50(left) and t=100(right), respectively.

When the Grashof number is low ($Gr = 5 \times 10^3$) the kinetic energies and streamlines plotted in the left part are in a good agreement. But, when Gr is high enough, e.g. $Gr = 5 \times 10^5$, the schemes produce two different types of results. The central difference scheme gives an oscillatory solution whereas those obtained by the other three schemes are steady. The central difference scheme, compared with the finite element method[16] and the Tau-Chebyshev pseudospectral method[17], is valid for the time-dependent simulation of the present study. The difference among the results produced by the four schemes when Gr is high is a consequence of diffusive-type errors on the discretization.[19] The results for the different schemes should be the same if calculations are taken on a finer grid. Bottaro and Zebib[20] gave a detailed discussion on the problem of discretization and showed that a designed coarser grid, when the central difference scheme is used, provides the solutions agreeing well with those obtained by a very fine grid.

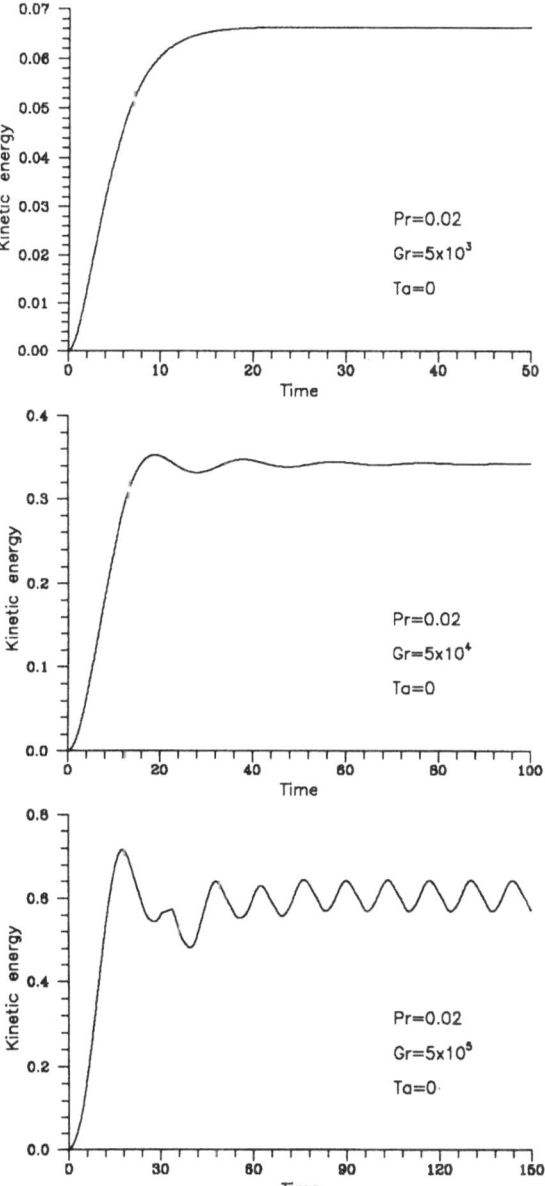

Figure 7. The time evolution of the kinetic energy K for Gr= (a)5×10^3, (b)5×10^4, and (c)5×10^5 at Ta=0 with Pr=0.02.

The unsteady behavior of melt flow at Ta=0 is illustrated by the time evolution of the kinetic energy K. Figure 7 presents them for three Gr, 5×10^3, 5×10^4 and 5×10^5. In the case of Gr $= 5 \times 10^3$, K approaches a steady value. From Figure 6 we see only one convective roll in the boat. When Gr reaches 5×10^4, a moderate value, K implies a dampened oscillatory solution. If Gr increases high enough, e.g. at 5×10^5, K no longer converges but is periodic after several cycles. The flow at Gr $= 5 \times 10^5$ is oscillatory. We plot the streamlines and isotherms at several typical times in Figure 8. At first the melt rises at the hot wall, moves from left to right, and descends at the cold wall. After that the flow structure becomes quite complex. The primary convective roll is not a single one but presents patterns of two or three co-rotating vortices. Meanwhile, some small eddies are generated at the bottom of the boat. During one period of oscillation the primary convective

75

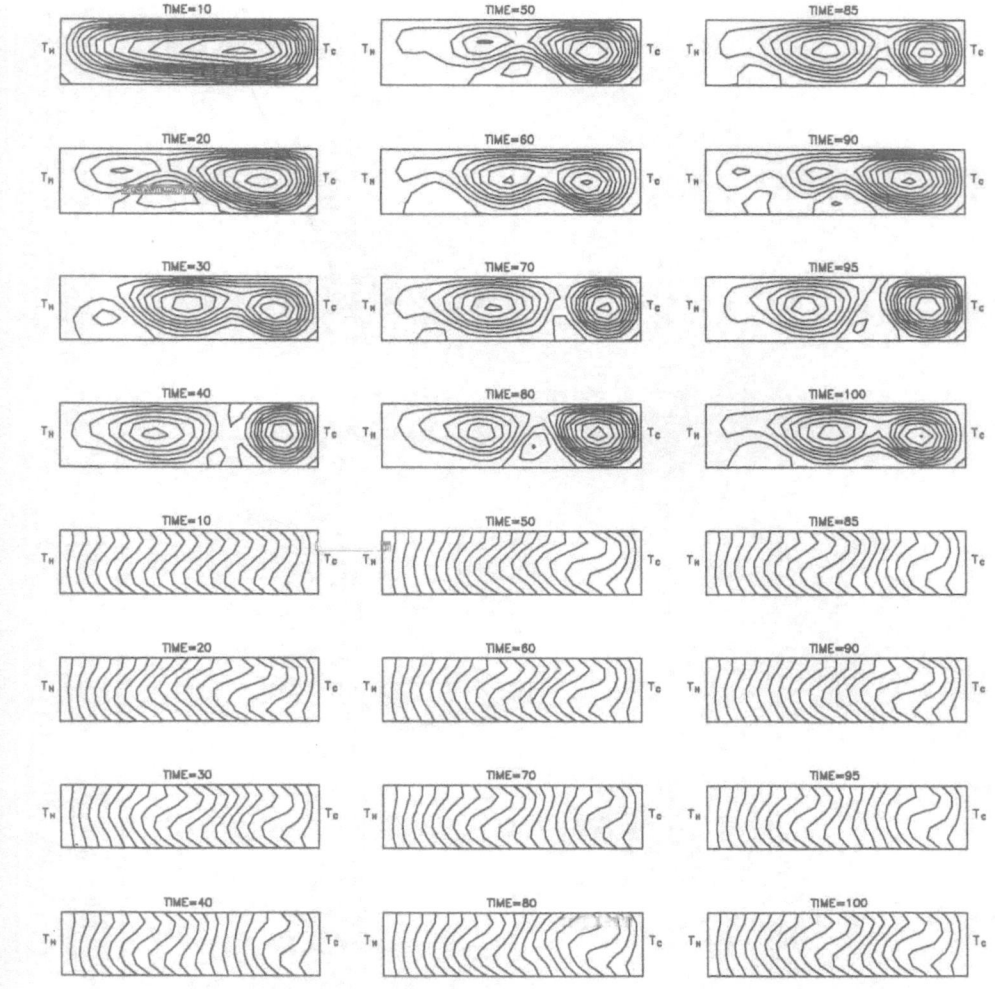

Figure 8. Streamlines and isothermal at several times for $Gr=5 \times 10^5$, Ta=0 and Pr=0.02.

roll is divided or merged accompanied by the growth or decay of these small eddies.

The melt flow at Ta=0 can be considered gravity-dependent via Gr, when the temperature difference between the ends of the boat is held constant. From the above results we see that centrifugal acceleration may enhance the strength of convection and induce oscillation in the melt but does not make stabilize the melt flow.

In the following, we take the Coriolis force into account. To evaluate the Coriolis effect on the melt flow, especially on oscillatory convection, the case at $Gr = 5 \times 10^5$ is considered. The simulations were performed for a variety of |Ta|, from 1×10^4 to 1×10^9. When |Ta| is less than 1×10^8, no difference between these results and those at Ta=0 can be observed. However, when |Ta| reaches 1×10^9 the results shown in Figure 9 are much more interesting. Figure 9(a) shows the time evolution of the kinetic energy K for $Gr = 5 \times 10^5$ and $Ta = 1 \times 10^9$. According to the definition 14 the centrifuge rotates counterclockwise if $Ta = 1 \times 10^9$. Comparing with Figure 7(c), the melt flow still maintains a sustained periodic oscillation. The strength of the convection seems to be enhanced by the Coriolis force. When the centrifuge runs in clockwise direction, i.e. $Ta = -1 \times 10^9$, the result shown in

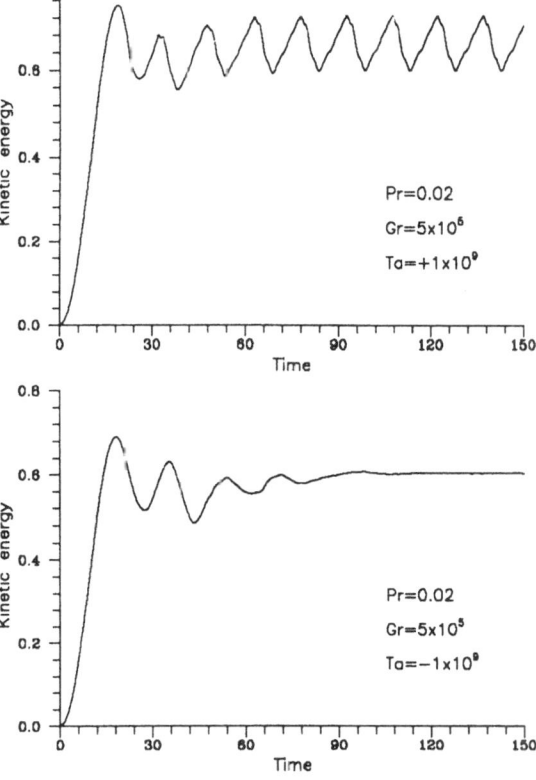

Figure 9. The time evolution of the kinetic energy K for $Gr=5 \times 10^5$, $Ta=$ (a)1×10^9 and (b)-1×10^9, and Pr=0.02.

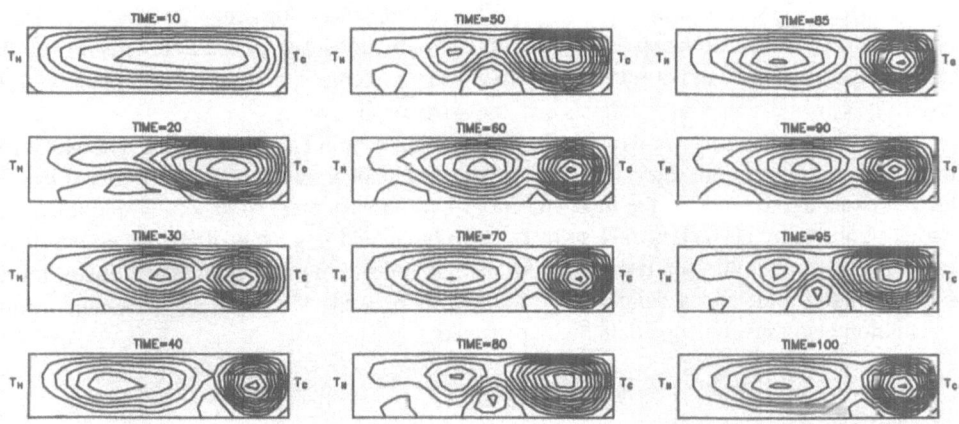

Figure 10. Streamlines at several times for $Gr=5 \times 10^5$, $Ta=1 \times 10^9$ and Pr=0.02.

Figure 9(b) approaches a steady value, and the melt flow is not oscillatory, but steady.

The results lead to a reexamination of stabilizing effects due to the Coriolis force observed in our model experiments[15]. It must be noted that the Coriolis force does not affect the melt flow by doing work. It influences the melt flow by altering the flow pattern, and then the balance of the system energy. In order to analyze the Coriolis effect on the flow structure, we plot the streamlines in Figure 10 for $Ta = 1 \times 10^9$ at several typical times and in Figure 11 for $Ta = -1 \times 10^9$ at one time when the flow becomes steady. The isotherms for these two cases are nearly the same as those at Ta=0 and not presented here. The calculations demonstrate that the Coriolis effect on the melt is not so easy as shown in Figure 3. Both cases (I and II) may coexist in the flow regimes. Analysis reveals that, for $Ta = 1 \times 10^9$, the primary convective rolls tend to be cut by the Coriolis force when it helps the small eddies to grow. This indicates that the flow could not become steady but more oscillatory. On the contrary, the effects of the Coriolis force are opposite completely if $Ta = -1 \times 10^9$. This time the Coriolis force expands the primary vortex but prevents the growth of the small eddies. This is perhaps the reason why the steady state of the melt does occur.

TIME=100

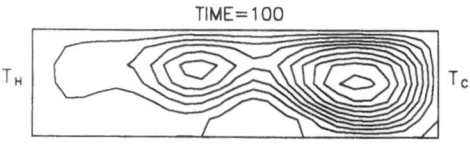

Figure 11. Streamlines for $Gr=5 \times 10^5$, $Ta=-1 \times 10^9$ and Pr=0.02 at t=100, at which the melt flow has become steady.

CONCLUSIONS

A numerical investigation was carried out for melt flow in a simplified two-dimensional rectangular boat influenced by both high gravity (i.e. the centrifugal acceleration) and the Coriolis force. The results show that the Coriolis force may have a stabilizing influence on melt flow but only under the conditions of co-rotation of the centrifuge and the main fluid flow of the melt within a certain range of Taylor number. When the Coriolis force is not involved the same stabilizing effect of high gravity does not occur. The results in this numerical study verify the observation of model experiments[15], and partially explain why the striations of the grown GaAs crystals become weak or fade away..

In the present study we assume fixed temperature boundary conditions. The isotherms move only slightly for all the simulations. However, in all actual processes of crystal growth heat is exchanged between the melt and its growth environment, and the temperature field varies significantly. Thus a further study of the influence of large centrifugal acceleration and the Coriolis force on crystal growth, espeacially in considering radiative heat transfer on the surface of the melt and a solid-liquid interface, is needed. Moreover, a three-dimensional calculation also must be considered indispensable.

ACKNOWLEDGMENTS

The present work was initially partly supported by the National Natural Sciences Foundation of China and subsequently by the Chinese Academy of Sciences. The simulations was performed on Alliant FX/40 computer of the institute of Mechanics. The two first authors wish to acknowledge youth research scholarships from the Institute of Mechanics. Also we would like to thank Professor G. Müller and Mr. J. Friedrich of Universität Erlangen-Nürnberg for their valuable comments and suggestions concerning this paper.

REFERENCES

1. J.R. Carruthers, *J. Crystal Growth* 32:13 (1976).
2. W.E. Langlois, *Ann. Rev. Fluid Mech.* 17:191 (1985).
3. G. Müller, "Crystal Growth from the Melt," Springer, Berlin (1988).
4. G. Müller, P. Kyr and E. Schmidt, *J. Crystal Growth* 49:387 (1980).
5. G. Müller, G. Neumann, *J. Crystal Growth* 59:548 (1982).
6. G. Müller, G. Neumann, *J. Crystal Growth* 63:58 (1983).
7. G. Müller, *J. Crystal Growth* 99:1242 (1990).
8. W. Weber, G. Neumann and G. Müller, *J. Crystal Growth* 100:145 (1990).
9. G. Müller, G. Neumann and W. Weber, *J. Crystal Growth* 119:8 (1992).
10. H. Rodot and L.L. Regel, *J. Crystal Growth* 79:77 (1986).
11. H. Rodot and L.L. Regel, *J. Crystal Growth* 104:280 (1990).
12. X.R. Zhong, B.J. Zhou, Q.M. Yan, F.N. Cao, C.J. Li, L.Y. Lin and W.J. Ma, Y. Zheng, F. Tao, M.L. Xue, *J. Crystal Growth* 119:74 (1992).
13. B.J. Zhou, F.N. Cao, L.Y. Lin and W.J. Ma, Y. Zheng, F. Tao, M.L. Xue, Growth of GaAs single crystals under high gravity conditions, "The 2nd International Workshop Materials Processing in High Gravity," Potsdam, New York, (June 1993).
14. W.A. Arnold, W.R. Wilcox, F. Carlson, A. Chait and L.L. Regel, *J. Crystal Growth* 119:24 (1992).
15. W.J. Ma, F. Tao, Y. Zheng, M.L. Xue and B.J. Zhou, L.Y. Lin, Response of temperature oscillations in a Tin melt to the effects of centrifuge, "The 2nd International Workshop Materials Processing in High Gravity," Potsdam, New York, (June 1993).
16. M.J. Crochet, F.T. Geyling and J.J. Van Schaftingen, Finite element method for calculating the horizontal Bridgman growth of semiconductor crystals, in: "Finite Elements in Fluids-Volume 6," R.H. Gallagher, G.F. Garey, J.T. Oden and O.C. Zienkiewicz, ed., John Wiley & Sons, Chichester (1985).
17. J.P. Fontaine, E. Crespo del Arco, A. Randriamampianina, G.P. Extrémet and P. Bontoux, *Adv. Space Res.* 8:265 (1988).
18. S.V. Patankar. "Numerical Heat Transfer and Fluid Flow," Hemisphere, Washington, D.C. (1980).
19. W.J. Minkowycz, E.M. Sparrow, G.E. Schneider and R.H. Pletcher, "Handbook of Numerical Heat Transfer," John Wiley & Sons, New York (1988).
20. A. Bottaro and A. Zebib, *Phys. Fluids*, 31:495 (1988).

VARIATION OF EFFECTIVE IMPURITY SEGREGATION COEFFICIENT IN TELLURIUM GROWN UNDER HIGH GRAVITY

I.I. Farbshtein,[1] R.V. Parfeniev,[1] N.K. Shulga[1] and L.L. Regel[2]

[1]A.F. Ioffe Physico-Technical Institute
Russian Academy of Sciences
St. Petersburg, Russia
[2]International Center for Gravity Materials Science and
Applications
Clarkson University
Potsdam, NY 13699-5700 USA

ABSTRACT

Previous experimental results on tellurium crystal growth in a Bridgman configuration under normal and enhanced gravity (5 g, 10 g) were analyzed using a thermosolutal convection model. For the given processing conditions the calculated Rayleigh numbers are in the range for stable convection in the melt. We separated the contributions of impurity and intrinsic defects using the experimental electrical properties along the ingots. The shape of the charge carrier profile depended on the relationship between the impurity con-centration and the number of electrically active defects. It was shown that with steady convection, the gravity level affects the dynamics of intrinsic defect formation. The Hole mobility in ingots solidified at high gravity was lower than in the ingot solidified at 1 g.

INTRODUCTION

The objectives in growing crystals are mostly determined by applications of the crystals as basic materials for devices (electronic, optical, etc.) and for scientific investigations (solid state physics, crystallography, etc.). All these applications need crystals that must be grown reproducibly with a well defined size, shape, and orientation. But real crystals are imperfect in most cases, limiting their applicability or profitability of their production. The improvement of the crystal quality and the growth process demands a deeper under-standing of the correlation between the formation of crystal imperfections and growth conditions. The action of gravity is one of these conditions; though 20 years ago it was considered invariable and natural. At present, spacecraft and centrifuges offer acceleration over a wide range, between one millionth of g (g = 9.81 m/s^2) and several million g. This paper contains a further analysis of the experimental results on tellurium crystal growth under high gravity (5 g and 10 g) performed in 1985.[1]

Recent studies of crystal growth phenomena at different gravity levels demonstrate that the main way in which gravity acts on the growth of crystals is through bouyancy-driven convection.[2] In this paper we consider convection during the directional solidification of molten tellurium, and the resulting distribution of both electrically active impurities and intrinsic imperfections in the crystal.

EXPERIMENTS

As was reported earlier,[1] the Te crystals were grown by the gradient freeze technique using a single crystal Te seed, a furnace with a temperature gradient of 30 °C/cm, and controlled cooling. The seed had been prepared by cleaving a Te single crystal along the threefold axis (the [0001]-direction), which became the growth direction of the Te crystals. The feed material was high purity tellurium with a hole concentration near 10^{14} cm^{-3} at 77 K due to traces of Sb. Melting of the feed ingot was followed by homogenization for 2 hours at 500 °C during rotation of the centrifuge. After that, the solidification was begun with a cooling rate of ~ 75 °C/h. When the furnace temperature had been lowered to 350 °C, the centrifuge was switched off and the crystals were annealed for 2 hours at this temperature.

This configuration is considered thermally stabilized, because the crystal is situated under the melt, and the direction of the acceleration vector is opposite to the direction of solid-melt interface movement. The geometrical and thermal parameters of the growth system under consideration are listed in Table 1.

Table 1. Geometrical and thermal parameters of the growth system.

Parameter	Value
Height of the melt H	45 mm
Ampoule diameter D	6.5 mm
H/D	~ 7
Temperature of cool zone, T_c	450 °C
Temperature of hot zone, T_h	600 °C
Gradient zone length, L_g	45 mm

DISCUSSION

THE CONVECTIVE FLOW REGIME

An ideal vertically stabilized configuration would not have any buoyancy flows if the melt density decreased with temperature. But, in fact, lateral temperature gradients always exist and cause lateral density variations in the liquid. The radial temperature gradient may be taken as proportional to the axial gradient in the gradient zone $(T_h-T_c)/L_g$, where T_h and T_c are the temperatures of hot and cool zones, respectively, and L_g is the adiabatic zone length. We use as a measure of the buoyancy driving force the radial Rayleigh number, as done by Chang and Brown:[5]

$$Ra_r = \frac{\beta\, g\left[(T_h - T_c)/L_g\right]R^4}{\alpha\, v} \tag{1}$$

where β is the thermal expansion coefficient, g is acceleration, α is the thermal diffusivity, and v is the kinematic viscosity of the melt. The properties of molten tellurium are given in Table 2.

Table 2. Properties of molten tellurium.

Properties	Value
Density of the melt at $T_{m.p.}$	5700 kg/m^3
Thermal conductivity at 500 °C	3.0 W/m.K
Specific heat, C_p	295 J/kg.K
Thermal diffusivity, $\alpha = k/C_p\,\rho$	1.7×10^{-6} m^2/s
Thermal expansion coefficient at $T_{m.p.}$, β	3.72×10^{-4} K^{-1}
Prandtl number Pr$=v/\alpha$ at 500 °C	0.7

Molten tellurium near the melting point ($T_{m.p.}$=452 °C) has two peculiar features associated with its quasi-polymer chain structure that are not destroyed immediately upon melting:

(a) **Density anomaly.** Liquid Te has a smooth minimum of density near $T_{m.p.}$, where the average number of the first neighbors is near 2 (2.5 at 490 °C). That is, near the melting point many atoms have only two nearest neighbors, because they are covalently bounded in the chain.[4] When the temperature is raised from 490 °C to 930 °C, the average coordination number rises from 2.5 to 3. This high temperature structure of liquid tellurium can be represented as a three-dimensional network of covalent bonds, joining each atom with three nearest neighbors. It is an arsenic (A7) type of structure.

(b) **Viscosity anomaly.** The molten tellurium is an associated liquid, as pointed out by Cabane and Friedel,[4] with long-range order partially remaining near the melting point.

Therefore, its viscosity near the melting point is one order of magnitude larger than liquid metals. On heating, the viscosity decreases to normal values.

These two anomalies act in different ways on natural convection in molten Te. The former makes the system thermally unstable in the region 450-500 °C. In the present experiments, this corresponds to ~ 10 mm height of melt near the solid-liquid interface. This density inversion should lead to an additional convective transfer. But at the same time, the high viscosity should suppress buoyancy flows in roughly the same temperature range (450-500 °C). The viscosity dependence of the Rayleigh number for different gravity levels is defined in Table 3. Supposing the former effect to be compensated approximately by the latter one, we can consider the convection in our system as existing throughout the melt volume in spite of the high viscosity of the melt near the bottom.

Table 3. Rayleigh number calculated for molten tellurium at different accelerations.

T (°C)	$v = f(T)$ (cm^2/s)	Ra		
		1 g	5 g	10 g
452	10^{-3}	2	10	20
460	2.7×10^{-6}	3.2×10^2	1.6×10^3	3.2×10^3
500	1.2×10^{-6}	0.7×10^3	3.5×10^3	7.0×10^3
510	6.06×10^{-7}	1.4×10^3	7.0×10^3	1.4×10^4
530	2.7×10^{-7}	3.0×10^3	1.5×10^4	3.0×10^4
550	2.4×10^{-7}	3.6×10^3	1.8×10^4	3.6×10^4
570	2.23×10^{-7}	4.0×10^3	2.0×10^4	4.0×10^4
600	1.83×10^{-7}	4.8×10^3	2.4×10^4	4.8×10^4

To evaluate the convective regime in our system, we used the critical Rayleigh numbers generalized by Müller[5] for semiconductor crystal growth. The following values were used for our geometric parameters (Table 2):

(a) Critical Rayleigh number for the transition from pure diffusive transport to laminar convective transport, $Ra^{c1} = 10^3$.

(b) Critical Rayleigh number for the transition from laminar to turbulent convection, $Ra^{c2} = 10^6$.

It is seen that our values of Ra lie between Ra^{c1} and Ra^{c2}, which means that in our growth system we deal with steady convective heat and mass transport. The absence of dopant striations in the samples is further evidence that the convection was laminar and not time dependent.

Concentration and Hall Mobility of Hole Charge Carriers

Due to the comparatively large viscosity of molten Te, rearrangement of the atoms is possible even after the temperature falls below the melting point. This small displacement of atoms after solidification may be increased by the more vigorous convection caused by increased gravity, leading to the formation of lattice defects. Such a structure rearrangement is weak, but with increase of the melt viscosity it may be great enough for the long-range order to be partially destroyed, as observed in the Te-Se alloys, where the addition of selenium increases the viscosity and disturbs the long-range order. The greater the viscosity (or Se-content), the stronger the disturbance in the solidifying melt. Thus more intensive convection near the solid-melt interface may displace the atoms slightly and generate additional defects.

Crystalline tellurium is a p-type semiconductor in the extrinsic conduction region. The hole concentration is about 10^{13} cm^{-3}, which seems to be a lower limit and may be due to lattice defects. These defects have an acceptor behavior and their nature is not yet clear.[6]

The electrical activity of impurities in Te was considered by Klitzing.[7] Since the concentration, p, of holes in crystalline Te in the extrinsic region is an integral result of both intrinsic lattice defects and electrically active dopants, it is important to separate their contributions to the hole concentration. This can be done by comparison of the variations of Hall mobility, U_H, and hole concentration, p, along the samples. The variation of U_H down the ingot is shown in Fig. 1 for the samples grown at 1, 5, and 10 g. For steady buoyancy-driven convection, which was established above, the longitudinal macroscopic impurity segregation should be described by the Pfann equation:[8]

$$C = k\,C_0\,(1 - z)^{k-1} \quad , \tag{2}$$

where k is the impurity segregation coefficient, C_0 is the initial impurity concentration, and z is the fractional distance down the ingot. The data are plotted as log(p) versus log(1-z) in Fig. 2.

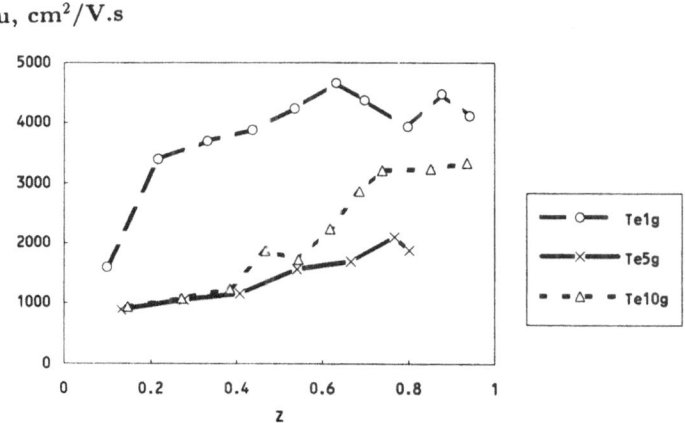

Figure 1. Hall mobility versus fraction solidified Te (T=77.4 K).

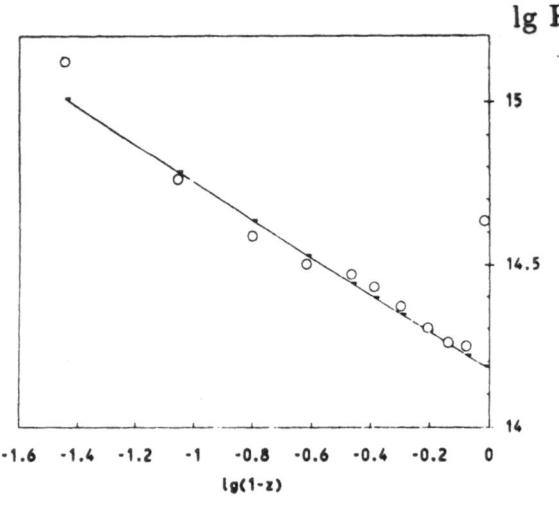

a) 1g

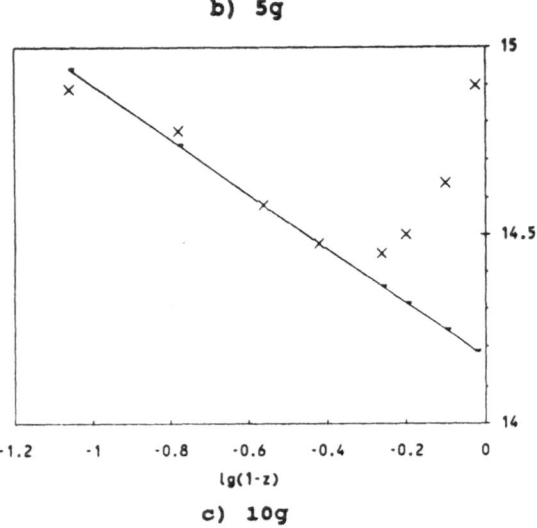

b) 5g

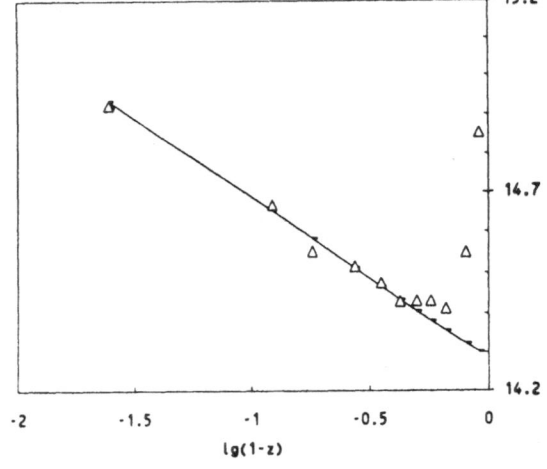

c) 10g

Figure 2. Hole concentration, p, in Te samples solidified at (a) 1 g, (b) 5 g, and (c) 10 g. Data points and correlating lines.

As mentioned above, the hole concentration, p, is a result of the addition of two processes. The first process is the macrosegregation of a dopant, and the second one is a drastic fall of the number of defects at the beginning of crystallization. The decrease of the defect concentration leads to a reduction of hole concentration produced by lattice imperfections and improvement of the crystal structure. This explanation of the high hole concentration near the seed is confirmed by the low mobility in this region, and the gradual increase of U_H along the samples (Fig. 1). Such a behavior of U_H is characteristic of crystals grown at 5 and 10 g; but, in the analogous sample obtained at normal gravity U_H has a constant value of ~ 3000 cm^2/V.s.

If the measured hole concentration p is the sum of the concentrations of intrinsic defects, C_{def}, and electrical active impurity, C_{im}, then $p = C_{def} + C_{im}$. For $C_{def} \ll C_{im}$, the effective segregation coefficient can be determined when $k = C_{im}/C_L < 1$. This situation is observed in the samples for $z > 0.6$ (5 g and 10 g) and $z > 0.2$ (1 g), where the maximal mobility occurred.

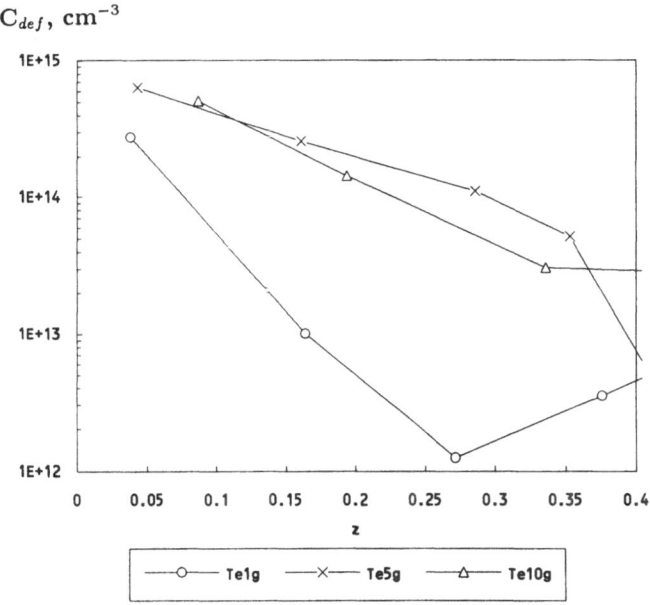

Figure 3. Defect concentration following the seed. (Difference between experimental points and correlating lines in Fig. 2.)

The plots of log(p) versus log(1-z) in Fig. 2 show linear behavior, with a slope corresponding to the following effective segregation coefficients: $k = 0.42$ for 1 g, $k = 0.27$ for 5 g, and $k = 0.60$ for 10 g. By extrapolating the straight lines to the initial parts of the ingots, we obtain the defect distribution as a difference between the measured hole concentration and the extrapolated impurity segregation. The data resulting from this procedure are shown in Fig. 3. The fall of the defect concentration in the initial part of the samples ($z < 0.3$) has an exponential behavior:

$$C_{def} = C_{def}^0 \exp(-z/z_0) \qquad (3)$$

The initial concentration, C^0_{def} and the relaxation length, z_0, for the samples are:

1 g : $C^0_{def} = 6 \times 10^{14}$ cm^{-3}, $z_0 = 0.05$;

5 g : $C^0_{def} = 8.4 \times 10^{14}$ cm^{-3}, $z_0 = 0.14$;

10 g : $C^0_{def} = 1.3 \times 10^{15}$ cm^{-3}, $z_0 = 0.09$.

These results show that enhanced gravity gives rise to additional defects with greater z_0 in the Te crystals.

The shape of the $p(z)$ curve depends on the relation between the impurity and defect concentrations. The most typical case is for very pure Te crystals, where $C_{def} > C_{im}$. In this case, the hole concentration distribution along a crystal may correspond to Eq. (2) nowhere, as observed in our previous experiment on the Te polycrystalline sample obtained at 5 g.[9] The minimum of p may be displaced along the z-axis due to the change of the defect relaxation rate (i.e. the change of z_0 in Eq. (3)). Moreover, in the part of the sample where $C_{def} < C_{im}$, a local increase of C_{def} may occur on a grain boundary, leading to violation of Eq. (2). Thus, for a pure polycrystalline Te sample, it is difficult to say anything about the segregation coefficient of an impurity.

CONCLUSION

We have considered data on the segregation of traces of Sb in tellurium crystals grown at enhanced gravity (5 g and 10 g). These data were obtained by means of Hall effect measurements with extrinsic conductivity, which is the only way to determine the content of an electrically active dopant in semiconductors with $< 10^{-3}$ at.% of impurities.

In order to reveal the impurity distribution, it is important to take into account the contribution of intrinsic defects to the concentration of charge carriers; because not only impurities produce holes in Te, but so do defects.

As mentioned above, the minimum concentration of defects in Te is $(0.5-1) \times 10^{14}$ cm^{-3}, which limits the applicability of this method. Separating the partial contribution of electrically active defects and impurities to the hole concentration, the effective segregation coefficients and the distribution of defects along Te samples were determined for three gravity levels: 1, 5, and 10 g.

REFERENCES

1. L.L. Regel, A.M. Turchaninov, R.V. Parfeniev, I.I. Farbshtein, N.K. Shulga, S.V. Nikitin and S.V. Yakimov, *J. Phys. III France* 2:373 (1992).
2. G. Müller, *J. Cryst. Growth* 99:1242 (1990).
3. C. Chang and R. Brown, *J. Cryst. Growth* 63:343 (1983).
4. B. Cabane and J. Friedel, *J. de Phys. France* 32:73 (1971).
5. G. Müller, Convection and inhomogeneities, *in*: "Crystal Growth from the Melt," H.C. Freyhardt, ed., Springer-Verlag, Berlin (1988), pg. 143.
6. W. Baier and H. Kohler, *J. Phys.: Condens. Matter* 3:307 (1991).
7. K. Klitzing, *Solid State Electron.* 21:223 (1978).
8. W.G. Pfann, *Trans. AIME* 194:747 (1952).
9. L.L. Regel, R.V. Parfeniev, I.I. Farbshtein, N.K. Shulga, A.M. Turchaninov and B.T. Melech, "Proceedings of Congress of International Astronautical Federation, Austria," Pergamon Press, Oxford (1986).

ANALYSIS OF IMPURITY DISTRIBUTION BY GALVANOMAGNETIC METHOD IN InSb OBTAINED UNDER HIGH GRAVITY CONDITIONS

I.I. Farbshtein,[1] R.V. Parfeniev,[1] S.V. Yakimov,[1]
L.L. Regel,[2] Ramnath Derebail[2] and W.R. Wilcox[2]

[1]A.F. Ioffe Physico-Technical Institute
Russian Academy of Sciences
194021 St. Petersburg, Russia
[2]International Center for Gravity Materials Science and Applications
Clarkson University
Potsdam, New York 13699-5700, U.S.A.

ABSTRACT

Polycrystals of undoped and Te-doped InSb have been obtained under macrogravity conditions up to 10 g by directional solidification. The electroresistance and the Hall effect along the sample length were studied at 77-300 K to find out the role of gravitational forces in the crystallization processes.

INTRODUCTION

High gravity experiments can help to understand the role of gravity in semiconductor technology. InSb is a well-studied material and suitable for such experiments among the narrow gap semiconductors. Samples of InSb were obtained by the gradient freeze method under high gravity conditions (1-10 g). The morphologic characteristics of these samples were presented previously.[1] Each sample had a different polycrystalline structure. In this research we determined the dependence of carrier concentration on the distance from the start of crystallization as a function of acceleration. To determine impurity concentrations in relatively pure semiconductors, the galvanomagnetic method is preferable to analytical methods.

EXPERIMENT

Two series of samples were investigated: the first was Te-doped samples of InSb and the second was not deliberately doped. Hall voltages and electroresistivity voltages from 6 pairs of contacts were measured along the length of the ingot at room and liquid nitrogen temperatures. The type of the conductivity, the carrier concentration, and the mobility were determined from the experimental data.

EXPERIMENTAL RESULTS AND DISCUSSION

Doped Samples

The Te-doped samples were n-type in the whole temperature region, 77-300 K. The calculated results obtained from the liquid nitrogen temperature data are shown in Fig. 1 (dots connected by dashed lines). Since the material was doped only by one donor impurity, the formula for partial liquid mixing with constant segregation coefficient k can be used:[2]

$$C = kC_0 (1 - x/\ell)^{k-1} \qquad (1)$$

where C is the impurity concentration in the solid phase, C_0 is the initial impurity concentration in the melt, x is distance down the ingot and ℓ is the total length of the ingot. By means of the least squares method, the parameter k was fitted to the formula. The resulting segregation coefficient k(Te) was close to unity, which corresponds to the literature data. It increased from 0.9 at 1 g up to 1.1 at 5 g and 10 g. The calculated curves for these parameters are presented as the solid lines in Fig.1.

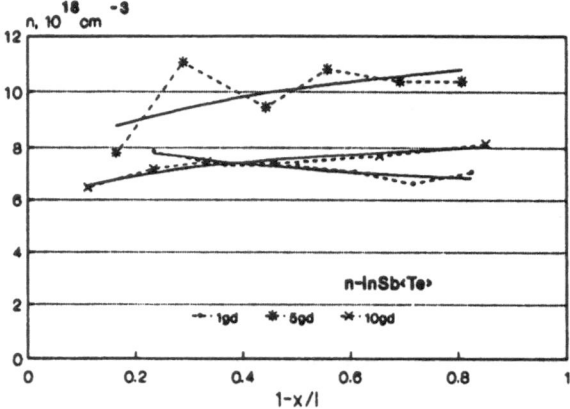

Figure 1. Donor impurity concentration n in Te-doped InSb ingots versus reduced distance. Experiments: dots connected by dashed lines. Correlation: solid lines, k = 0.9 for 1 g and 1.1 for 5 g and 10 g samples. T = 77.4 K.

It is interesting to note that the kind of segregation of Te in InSb changed with the acceleration: at 1 g the impurity was moved toward the end of the ingot (k(Te)<1), while at 5 g and 10 g it moved in the opposite direction (k(Te)>1). The maximum deviation of n from the correlation curve was at 5 g.

Undoped Samples

The distribution of the carrier concentration along the crystal length in ingots that were not deliberately doped is shown in Figs. 2 and 3 for 77 K (dots connected by dashed lines). The undoped samples were both n and p-type. They had not only a strong dependence of the measured concentration on x, but also a p-n junction at liquid nitrogen temperature in the ingot solidified at 7 g as well. These results indicate the simultaneous existence of two impurities in the samples: a donor and an acceptor, with the compensation degree changing along the ingots.

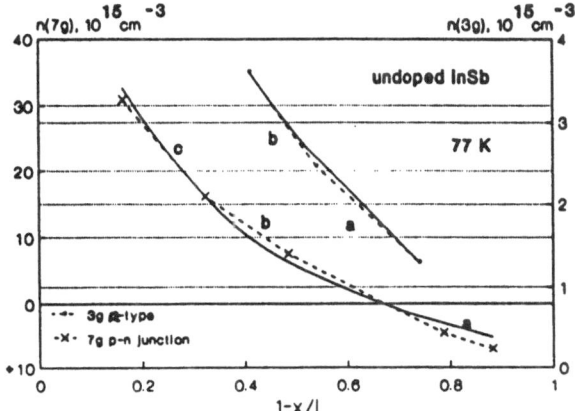

Figure 2. Difference C between donor and acceptor impurity concentrations in undoped strongly compensated samples verses reduced distance. Experimental: dots connected by dashed lines. Calculated curves: solid lines, $k_a = 1.01$, $C_{oa} = 511 \times 10^{15}$ cm^{-3}, $k_d = 0.55$, $C_{od} = 393 \times 10^{15}$ cm^{-3} for 3 g sample: $k_a = 0.92$, $C_{oa} = 64 \times 10^{15}$ cm^{-3}, $k_d = 0.65$, $C_{od} = 82 \times 10^{15}$ cm^{-3} for 7 g sample. T = 77.4 K.

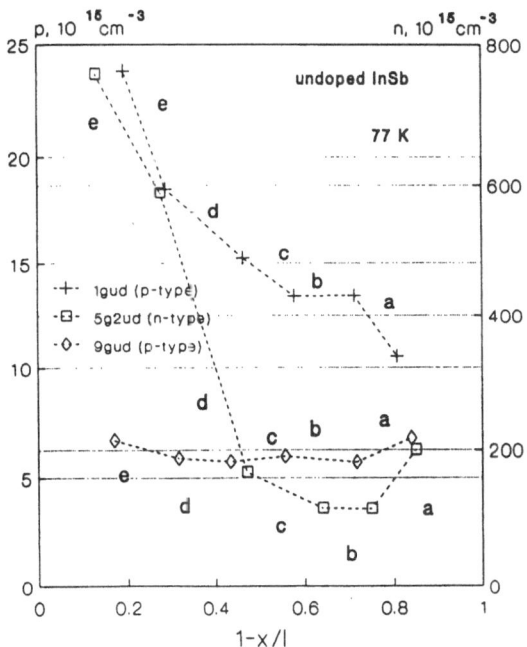

Figure 3. Difference C between donor and acceptor impurity concentrations in c undoped weakly compensated samples versus reduced distance for 1 g, 5.1 g, and 9 g. T = 77.4 K.

To estimate the segregation coefficient of an uncontrolled impurity in the samples, the slightly compensated regions were used. These regions were found by comparing the experimental carrier mobility data with the known dependence of mobility on carrier concentration in uncompensated n- and p-InSb (Figs. 4a and 4b). Here the solid lines denote the literature data, and the dots connected by dashed lines represent the present experimental results. Regions with the carrier mobility close to those of the undoped uncompensated samples existed in the samples obtained at 1 g (region b, c in Fig. 4b). In the slightly compensated regions, the carrier concentration profile at liquid nitrogen temperature was determined mostly by the distribution of only one impurity. In these cases is was possible to calculate the segregation coefficient using equation 1, yielding 1.01 for acceptors in the 1 g ingot, 0.65 for donors in the 5.2 g ingot and 0.92 for acceptors in the 9 g ingot.

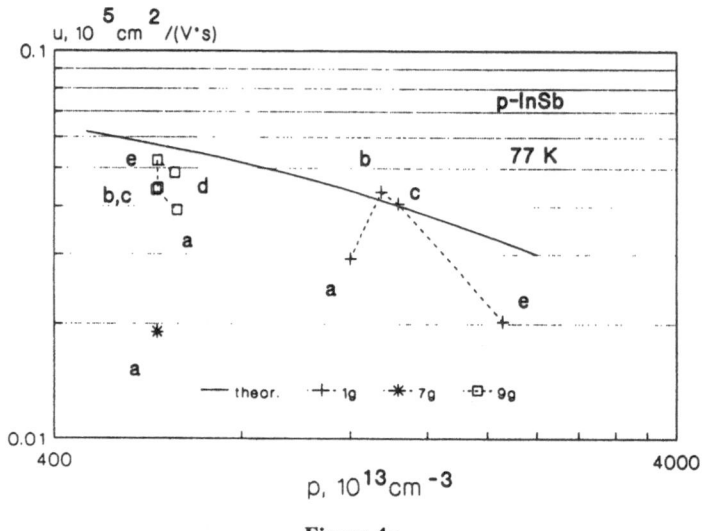

Figure 4a

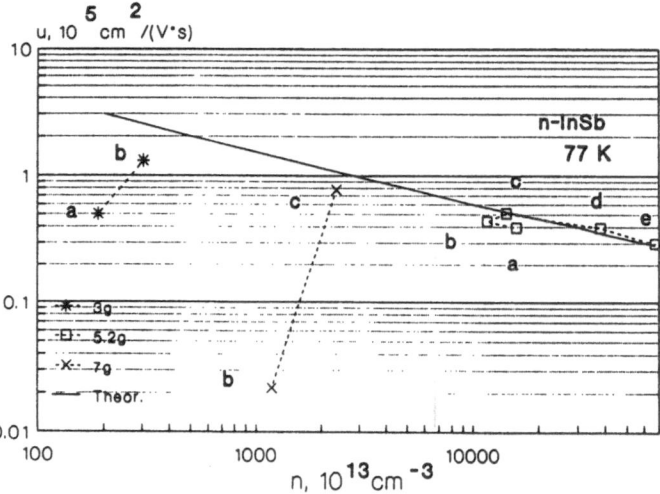

Figure 4b. Dependence of the carrier mobility u on the carrier concentration in (a) n-InSb and (b) p-InSb. The solid lines represent literature data. The dots connected by dashed lines are the present results for regions noted in Figs. 2 and 3 by letters. T = 77.4 K.

Slightly compensated regions were not found in the ingots obtained at 3 g and 7 g. Assuming the presence of two impurities (donor and acceptor) with independent segregation coefficients, an attempt was made to describe the net carrier concentration profile using the measured segregation coefficients in:

$$C = k_d C_{od} (-x/\ell)^{k_d-1} - k_a C_{oa} (1-x/\ell)^{k_a-1} \qquad (2)$$

where k_d and k_a are the segregation coefficients of donors and acceptors, C_{od} and C_{oa} are the initial concentrations of donors and acceptors in the melt, and C is the difference between the concentrations of donors and acceptors.

The best fit of the parameters was made to Eq. (2) by the least squares method. The resulting curves are in Fig. 2 by the solid lines. These show that the p-n junction obtained from the experiment can be described by Eq. (2) with C=O.

CONCLUSION

It was shown that the use of galvanomagnetic methods allows estimation of the segregation coefficient of donor and acceptor impurities both in compensated and uncompensated samples of InSb and their dependence on g.

A change of the type of segregation of Te in InSb with g was observed: a 1 g k(Te) was <1, and at 5 g and 10 g k(Te) was >1.

The influence of g on the segregation coefficient without deliberate doping was not found because of the strong sensitivity of the coefficient to freezing conditions, especially the melt-solid interface orientation. The Te segregation coefficient can vary from 0.47 to 4.2, the latter being for the (111) facet, i.e. up to 5.9-8.9 times.[3] In a polycrystalline material, the interface consists of a spectrum of crystal orientations. That is why measurements give different coefficients depending on the degree of polycrystallinity. Note that the dependence of C on the coordinate x in the ingots investigated was somewhat unmonotonic, especially for the 5 g doped sample, and was not described precisely by Eqs. (1) and (2) (Figs. 1 and 2). This behavior can be explained by the crystallization of several grains with different segregation coefficients.[4] In the investigated InSb there were many grains with different crystal orientations. The average grain was bigger in the samples obtained at higher g.[1]

REFERENCES

1.   R. Derebail, W. Wilcox and L. Regel, Directional solidification of InSb in a centrifuge, *J. Cryst. Growth* 119:98 (1992)
2.   W.G. Pfann, *J. Metals* 4:841 (1952).
3.   K.F. Hulme and J.B. Mullin, *Solid State Electr.* 5:211 (1962).
4.   J. Barthel, *Kristal und Technik* k77-k81 (1975).

THE INFLUENCE OF GRAVITY ON
Pb$_{1-x}$Sn$_x$Te CRYSTALS GROWN BY
THE VERTICAL BRIDGMAN METHOD

Y.A. Chen,[1] I.N. Bandeira,[1] A.H. Franzan,[1] S. Eleutério Filho,[2]
and M.R. Slomka[2]

[1]Laboratório Associado de Sensores e Materiais - LAS
Instituto de Pesquisas Espaciais - INPE
São José dos Campos, SP 12201, Brazil
[2]Centro Tecnológico de Informatica - CTI
Campinas, SP 13093, Brazil

ABSTRACT

Composition profiles of Pb$_{0.80}$Sn$_{0.20}$Te grown by the normal and inverted Bridgman methods are presented. The growth under stabilized solute gradient decreases the convection, and the final solute distribution corresponds to the partial diffusion-convective mechanism. Also shown are results of high-gravity gradient freeze growth with a destabilizing temperature gradient.

INTRODUCTION

The inverted vertical Bridgman method (IVB), where the solidification occurs from the top of the melt under a destabilizing thermal gradient, allows growth of crystals with buoyancy-driven convection different from that with the usual vertical Bridgman configuration (VB). Due to the unstable solutal gradient, the compositional profile of Pb$_{0.80}$Sn$_{0.20}$Te crystals grown by VB usually obeys the normal freezing law. It is well known that for Bridgman growth in a microgravity environment, the convection is suppressed, and a crystal with homogeneous composition can be grown with a diffusion-controlled steady state. However, experiments in space are restricted by technical and financial constraints, and alternatives, such as the IVB or growth under high gravity using centrifugal systems,[1-3] are becoming useful techniques to study the influence of gravity on crystallization processes.

Only a few materials have been grown by IVB.[4-7] Recently the pseudobinary compound

$Pb_{1-x}Sn_xTe$ has also been investigated.[8,9] This paper reports the influence of convection on the longitudinal composition profile in $Pb_{0.80}Sn_{0.20}Te$ crystals grown by VB and IVB at earth's gravity (1 g) and by the inverted gradient freeze technique at 3 g.

EXPERIMENTAL

The experimental system for inverted gradient freeze growth at 3 g is shown in Fig. 1. For all experiments the growth ampoule consists of a quartz rod of 8 mm diameter and 20 mm length with a $Pb_{0.80}Sn_{0.20}Te$ charge (8 mm diameter, 40 mm length, 69 purity) sealed in an evacuated (10^{-6} Torr) quartz ampoule with approximately the same internal diameter. This ampoule ($\sim$ 100 mm total length, 1.5 mm thickness) has a constriction ($\sim$ 10 mm length, $\sim$ 3 mm diameter) for seed selection. Enough charge powder is added to fill all empty spaces after melting. The quartz rod placed on top of the melt avoids material loss by evaporation. The quartz ampoule is sealed inside a steel cartridge.

For the VB and IVB experiments at 1 g, the cartridge containing the ampoule was placed in a vertical furnace with 80 cm length and 3.5 cm internal diameter. The furnace was heated to 950 °C, yielding a temperature gradient G of 16 °C/cm along the cartridge at the estimated position of the solid-liquid interface. Growth was caused by moving the

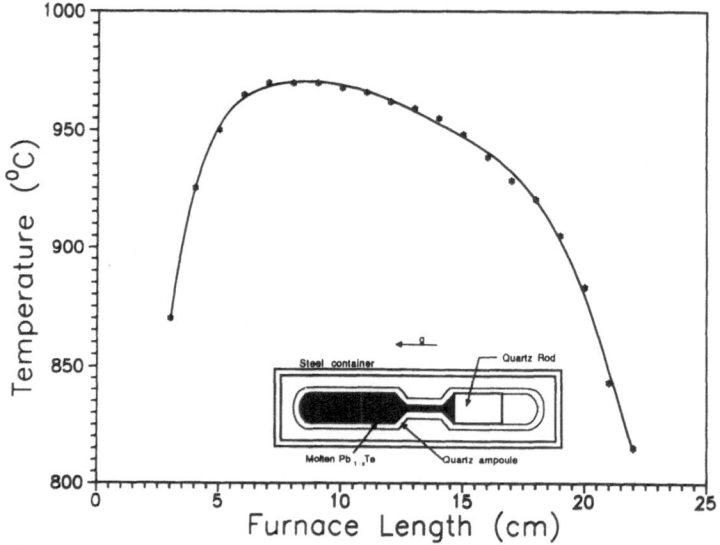

Figure 1. Temperature profile and ampoule arrangement for an inverted gradient freeze experiment at 3 g. The temperature profile was measured along the steel container.

ampoule through the furnace at V=1.35 mm/hr. For VB growth the ampoule was lowered by means of a stepping motor. For IVB growth the furnace was turned over and the ampoule slowly raised for growth.

The growth under high gravity was carried out in a smaller furnace (30 cm length and 2 cm internal diameter), which was attached to a centrifuge (maximum radius = 1 m) built at INPE. This centrifuge gives a resulting acceleration up to 10 g. After heating and stabilization, the centrifuge was accelerated to the desired level, and the growth was carried out by slowly lowering the furnace temperature. The growth conditions were approximately the same as at 1 g.

The SnTe axial concentration was measured through an Electron Probe Microanalyzer. The conversion from X-ray intensity to solute concentration was based on a calibration curve obtained from standard samples with known composition.

RESULTS AND DISCUSSION

During growth by VB the gravitational vector and temperature gradient are antiparallel. The solute SnTe is rejected from the solid-liquid interface, forming a piled-up layer on the liquid side. This accumulated layer with a less dense component is reduced by convective flows in the melt, and usually the composition of the growing material follows the normal distribution. Figure 2 shows the experimental compositions as points. The lines are third-order polynomial fits to the experimental data for mole fraction SnTe.

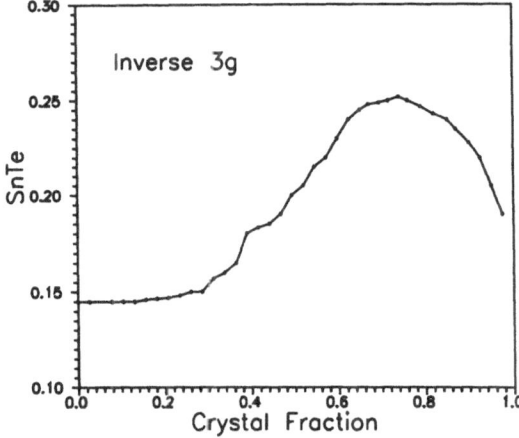

Figure 2. Solute distribution along the axial direction for crystals grown by the usual and inverted vertical Bridgman methods at 1 g.

In order to evaluate the influence of both solutal and thermal gradients on convection, their corresponding values of density gradient are compared. The density gradient caused by solute accumulation at the interface is given by:[10]

$$\left(\frac{\partial \rho}{\partial Z}\right)_C = \frac{\partial \rho}{\partial C}\frac{\partial C}{\partial Z} = \frac{\partial \rho}{\partial C}\left(\frac{1}{k_0} - 1\right)\frac{V C_0}{D} \tag{1}$$

at steady state. The density gradient due to the temperature gradient is:

$$\left(\frac{\partial \rho}{\partial Z}\right)_T = \frac{\partial \rho}{\partial T}\frac{\partial T}{\partial Z} = \frac{\partial \rho}{\partial T}G \tag{2}$$

Here $k_0 = 0.61$, $V = 1.35$ mm/h, $C_0 = 0.20$, $D = 5.3 \times 10^{-5}$ cm^2/s, and $G = 16$ °C/cm are equilibrium segregation coefficient, solidification rate, original mole fraction of SnTe, diffusion coefficient in the melt, and temperature gradient, respectively. The values $\partial\rho/\partial C$ = 1.62 g/cm^3.mol and $\partial\rho/\partial T = 8.3 \times 10^{-4}$ g/cm^3.°C are calculated from coefficients of bulk solutal and thermal expansion. Using these parameters, the numerical values of Eq. (1) and Eq. (2) are 1.46×10^{-1} g/cm^4 and 1.33×10^{-2} g/cm^4, respectively. The solutal density gradient is about ten times bigger than the thermal density gradient at the interface. This might explain the strong convective behavior presented in VB, and also the reduced convection during IVB growth (Fig. 2). However, since the solute concentration decreases exponentially into the liquid, it is likely that away from the interface, the temperature gradient becomes the dominant term. A complex compositional layering may occur in the melt.[11]

According to the results for IVB, the composition profile (Fig. 2) corresponds to partial melt mixing, where the convection flow is sufficiently weak to allow the formation of a solute boundary layer. Solutal convection is suppressed, because of the stable density gradient [Eq. (1)], and hence the diffusion-controlled solidification is not interrupted.

Convection is a function of the thermal Rayleigh number Ra$_T$, the solutal Rayleigh number Ra$_S$, the thermal Grashof number Gr$_T$, and the solutal Grashof number Gr$_S$. The definitions for Ra and Gr are:

$$\mathrm{Ra}_s = \frac{\beta_s}{Dv} gh^4 \frac{\partial C}{\partial z} \qquad \mathrm{Ra}_T = \frac{\beta_T}{\alpha v} gh^4 \frac{\partial T}{\partial z}$$

$$\mathrm{Gr}_s = g\beta_s h^3 \Delta C/v^2 \qquad \mathrm{Gr}_T = g\beta_T h^3 \Delta T/v^2$$

where β is the coefficient of bulk solutal or thermal expansion, D is the diffusion coefficient in the melt, v is the kinematic viscosity, α is the thermal diffusivity and h is the

Table 1. Rayleigh and Grashof numbers for Pb$_{0.80}$Sn$_{0.20}$Te.

h/ϕ	h(cm)	Ra$_T$	Ra$_S$	Gr$_T$	Gr$_S$	δ(cm)
0.625	0.5	2 x 10^3	5 x 10^6	2 x 10^4	2 x 10^5	3 x 10^{-2}
1.200	1.0	4 x 10^4	7 x 10^7	2 x 10^5	3 x 10^6	2 x 10^{-2}
2.000	2.0	6 x 10^5	1 x 10^9	6 x 10^6	4 x 10^7	1 x 10^{-2}
3.750	3.0	3 x 10^6	6 x 10^9	2 x 10^7	2 x 10^8	7 x 10^{-3}
5.000	4.0	1 x 10^7	2 x 10^{10}	6 x 10^7	7 x 10^8	5 x 10^{-3}

liquid height. Estimated values for our IVB experiments are shown in Table 1 for an ampoule diamter ϕ of 8 mm.

According to previous experiments,[8] the critical thermal Rayleigh number is 2×10^3 for an aspect ratio of 0.270, confirming that IVB was grown in the presence of some convection. The thickness of the solutal stagnant film δ can be estimated from a correlation for vertical zone melting of naphthalene:[12]

$$\delta \; = \; \frac{10r}{B\,(r/h)^{0.44}\,(h\,V/D)^{0.26}} \tag{3}$$

where $r = \phi/2$ is the ampoule radius, h is the height of the molten zone, and:

$$B \; = \; (Pr)^{-1/4}\,(Sc)^{1/2}\left[Gr_T + (Pr/Sc)^{1/2}\,Gr_S\right]^{1/4} \tag{4}$$

where $Pr = 2 \times 10^3$ (Prandtl) and $Sc = 53$ (Schmidt). Calculated values of δ are shown in Table 1. These yield estimates for $\delta V/D$ ranging from 0.004 to 0.02. Since $\delta V/D \ll 1$, the above correlation predicts complete mixing. This does not correspond to our experimental results. It should also be noted that the diffusional solute layer thickness D/V is approximately 13 mm.

The solute composition profile for inverted gradient freeze growth is shown in Fig. 3. This complex behavior is indicative of dramatic changes in convection or freezing rate as solidification proceeds.

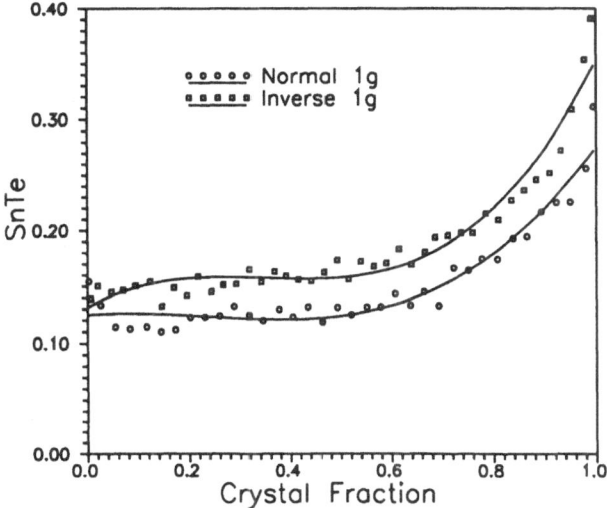

Figure 3. Solute distribution along axial direction for inverted gradient freeze growth at 3 g.

CONCLUSIONS

The inverted Bridgman method allows the growth of $Pb_{0.80}Sn_{0.20}Te$ crystals with a stabilizing solute gradient. Due to a reduction in convection, the final composition profile is similar to that provided by a partial diffusive mechanism. Inverted gradient freeze growth under high gravity yields a concentration profile indicating dramatic variation in convection or freezing rate during the solidification.

REFERENCES

1. G. Müller, E. Schmidt, and P. Kyr, *J. Cryst. Growth* 49:387 (1980).
2. H. Rodot, L.L. Regel, and A.M. Turtchaninov, *J. Cryst. Growth* 104:280 (1990).
3. L.L. Regel, A.M. Turtchaninov, O.V. Shumaev, I.N. Bandeira, Y.A. Chen, and P.H.O. Rappl, *J. Cryst. Growth* 119:94 (1992).
4. W.J. Boettinger, S.R. Coriell, F.S. Biancaniello, and M.R. Cordes. "NBS: Materials Measurements" (NBSIR 80-2082), Annual Report (July 1980).
5. G. Müller, G. Neumann, and W. Weber, *J. Cryst. Growth* 70:78 (1984).
6. K.M. Kim, A.F. Witt, and H.C. Gatos, *J. Electrochem. Soc.* 119:1218 (1972).
7. H. Jamgotchian, B. Billa, and L. Capella, *J. Cryst. Growth* 85:318 (1987).
8. K. Grasza and A. Jedrzejczak, *J. Cryst. Growth* 110:867 (1991).
9. K. Grasza and U. Zuzga-Grasza, *J. Cryst. Growth* 116:139 (1992).
10. R.M. Sharp and A. Hellawell, *J. Cryst. Growth* 12:261 (1972).
11. D.T.J. Hurle, G. Müller, and R. Nitsche, Fluid sciences and materials science in space, H.U. Walter, ed., European Space Agency, Springer-Verlag (1987).
12. W.R. Wilcox, Mass transfer in fractional solidification, *in:* "Fractional Solidification," M.Zief and W.R. Wilcox, eds., Dekker, N.Y. (1967) p. 61,62.
13. V. Fano, R. Pergolari, and L. Zanotti, *J. Mater. Sci.* 14:535 (1979).

MICROSTRUCTURAL DEVELOPMENT IN Pb-Sn ALLOYS SUBJECTED TO HIGH-GRAVITY DURING CONTROLLED DIRECTIONAL SOLIDIFICATION

R.N. Grugel,[1] A.B. Hmelo,[1] C.C. Battaile[2] and T.G. Wang[1]

[1]Center for Microgravity Research and Applications
Vanderbilt University
Box 6079-B, Nashville, TN 37235
[2]Department of Materials Science and Engineering
University of Michigan
Ann Arbor, MI 48109

ABSTRACT

Research conducted over the past three decades has suggested that solidification processing of metals and alloys in a centrifuge can lead to enhanced materials properties.[1-11] With the potential of such processing demonstrated, there exists a need for quantitative data gathered under controlled solidification conditions. To this end, a centrifuge, *dedicated to materials research*, was constructed within the Materials Science and Engineering Department at Vanderbilt University. This has since been employed to investigate the effect of a high-gravity environment on microstructural development of Pb- 50 wt pct Sn alloys during controlled directional solidification.

For otherwise <u>constant</u> solidification processing conditions of composition, growth rate, and temperature gradient, centrifugation caused the primary dendrite arm spacing to decrease significantly. The secondary dendrite arm spacing, the eutectic spacing, and the primary dendrite trunk diameters exhibited no change with increasing gravity level

These results are discussed in terms of suppressing convection in the bulk liquid and/or modification of the solute-enriched liquid layer about the dendrite tips. Work in progress to directly visualize effects attributed to enhanced gravity is discussed.

INTRODUCTION

It is well established that controlled directional solidification of metals and alloys can improve material properties of, for example, turbine blades.[12] Unfortunately, composition and/or temperature gradients in the liquid ahead of the interface can interact with Earth's gravity and initiate convection currents, which result in severe macrosegragation and, subsequently, inferior material properties.[13] Solidification processing in a centrifuge affords the opportunity of altering such convection patterns with, perhaps, the possibility of minimizing segregation and improving the microstructure.

A schematic representation of Vanderbilt's centrifuge facility is seen in Fig. 1. Briefly, the device consists of two 1.2 m arms mounted on a steel carriage that rotates about a vertical axis. The centrifuge is driven by a 1.5 hp DC motor and chain-and-sprocket assembly, which provides a maximum rotation rate of 120 rpm and a corresponding maximum acceleration of ~20 g. Thirty sliding electrical contacts and two rotary fluid lines are available to transfer power, thermocouple signals, and cooling liquids to and from experimental packages attached to the arms. Complete specifications regarding its design and construction are given elsewhere.[14]

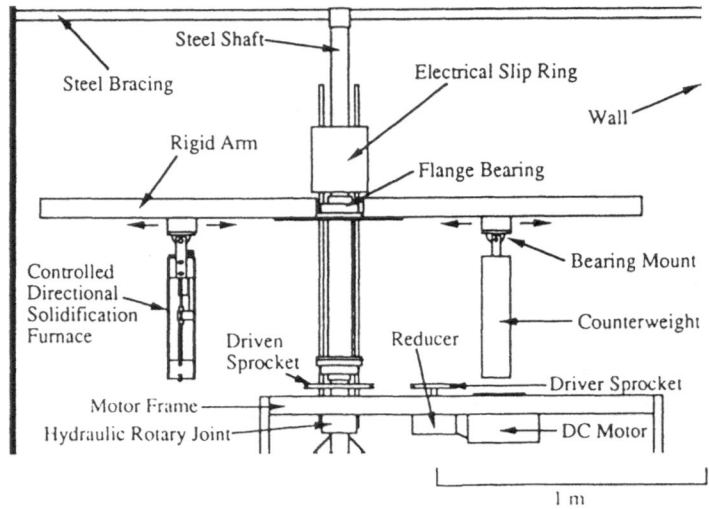

Figure 1. Schematic representation of the centrifuge facility.

The directional solidification furnace, Fig. 2, was bearing mounted onto the centrifuge to ensure alignment of the furnace axis with the resultant acceleration field. The assembly consists of a heating unit, a water-cooled toroid, and a sliding stage attached to a stepping motor, all mounted in an aluminum housing. More detail has been given previously.[14]

EXPERIMENTAL PROCEDURE AND RESULTS

The experimental procedure has been described in detail elsewhere.[14,15] Briefly, equal weights of tin (99.85%) and lead (99.9%) were melted together, thoroughly stirred, and cast into 5 mm ID quartz tubes. These were removed and remelted in closed-end 6 mm ID quartz tubes that contained fine K-type thermocouples. The sample was then placed into the furnace such that ~18 cm of the ~24 cm alloy length was remelted and allowed to come to equilibrium prior to being withdrawn at 21.1 $\mu m.s^{-1}$. The water-cooled toroid constrained the temperature gradient at the solid/liquid interface to 3.5 ± 0.5 K.mm^{-1}. Directional solidification continued for 6 cm, after which the sample was removed and quenched into water. Acceleration levels at the solid/liquid interface of 1.0, 2.5, 5.1, 7.1, 10.3, and 15.3 g were employed; three samples were processed at each level. In all cases the resultant acceleration vector was opposed to the growth direction.

The directionally solidified samples were sectioned and conventionally prepared for metallographic analysis with the primary dendrite arm spacing (λ_1), secondary dendrite arm spacing (λ_2), eutectic spacing (λ_E), and the primary dendrite trunk diameter (d) measured from each sample.

Figure 3 shows the measured secondary arm spacing λ_2, eutectic spacing λ_3, and primary dendrite trunk diameter d as a function of induced gravity level. The primary dendrite arm spacing λ_1 is shown in Fig. 4.

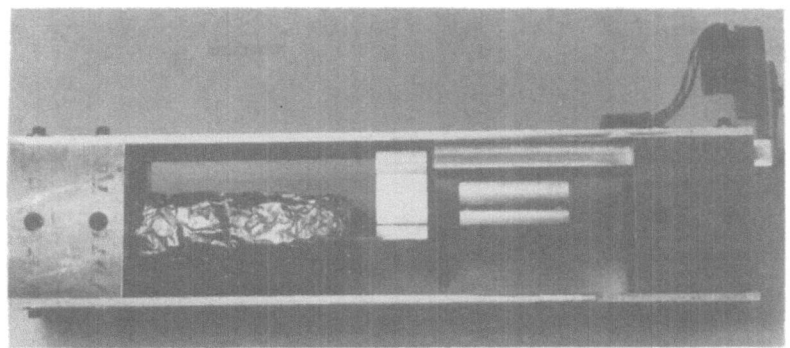

Figure 2. Photograph of the directional solidification apparatus.

DISCUSSION

The experimental results shown in Fig. 3 are predictable and well-characterized functions of the solidification processing conditions.[12,16] These did not significantly change as the imposed gravity level increased from 1 to 15.3 g, implying that centrifugation did not influence *interdendritic* solidification dynamics, Fig. 5 schematically depicts a dendritic array with the extent of the rejected solute field indicated.[17] Following the analysis of Allen and Hunt,[18] the solute diffusion fields of individual dendrites can be predicted to overlap ~250 µm behind the tips. Once past this transient region the concentration gradients are considerably smoothed. As the mushy zone extends ~3300 µm, the majority of interdendritic solidification and coarsening likely occurs in a stagnant and constrained, i.e. convectionless, environment.

In contrast to the results of Fig. 3, the primary dendrite arm spacing decreased with increasing gravity level, as shown in Fig. 4. Here the reduction in λ_1 is attributed to factors influencing the bulk liquid and/or the dendrite tip interface, i.e. diminished convection and/or altered mass transport.

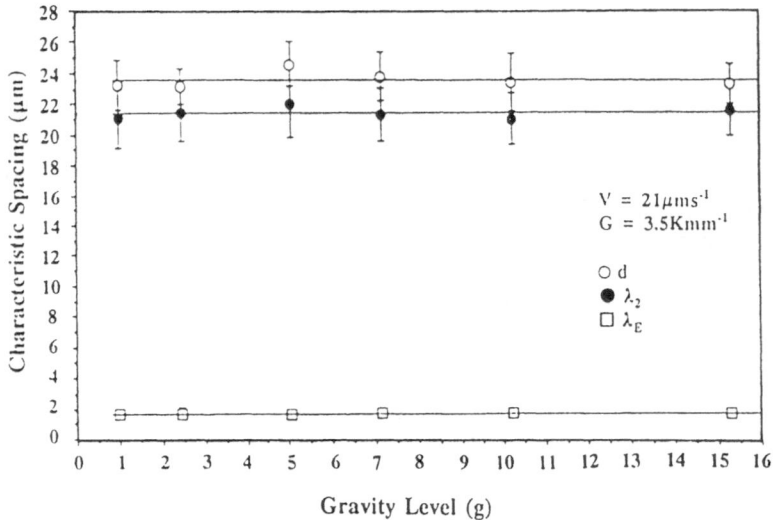

Figure 3. λ_2, λ_E, and d as a function of the imposed gravity level.

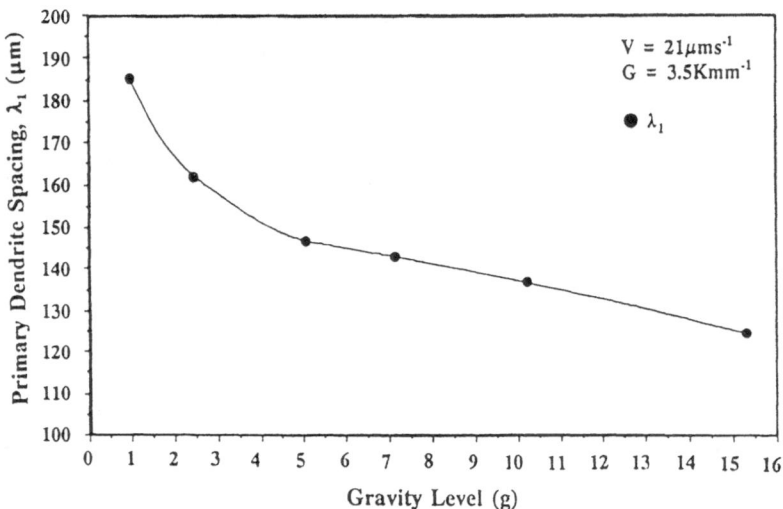

Figure 4. λ_1 as a function of the imposed gravity level.

Diminished Convection

The experimental and theoretical work of Müller *et al.*[2-4,6-8,11] suggests single-cell convective flows in the melt can be reversed by increasing the acceleration level. Rodot *et al.*[9,11] found that the distribution of Ag dopant in Bridgman-grown PbTe crystals improved at higher accelerations. It is also noted that the primary dendrite arm spacing in organic analogue materials increased when convection at the interface was initiated.[19,20]

Hunt[21] theoretically predicted the dependence of the dendrite arm spacing λ_1 on a number of solidification processing parameters and material properties, including the liquid diffusion coefficient. Using Liu's[22] properties for Pb-Sn alloys and the experimental values

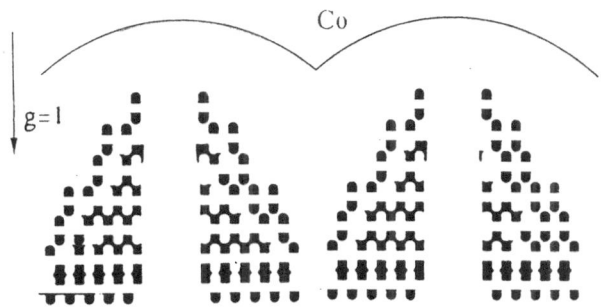

Figure 5. Schematic representationof the extent of the solute field about dendrites in unit-gravity.[17]

of G=3.5K.mm^{-1} and V=21.1μm.s^{-1}, a primary spacing for steady-state diffusion-controlled growth of $\lambda_1 \approx 160$ μm is predicted; this is comparable to the measured results for the high-gravity regime of Fig. 4. Frohberg *et al.*[23] reported that measured liquid self-diffusion coefficients for tin near the melting point can vary by as much as 50 to 100% or more between microgravity (convectionless) and unit-gravity environments. If a 50% increase is applied to the diffusion coefficient (D=717 μm^2.s^{-1}), the predicted primary spacing becomes $\lambda_1 \approx 180$ μm; a value which is in agreement with the spacings measured from 1 g experiments.

The above analysis suggests that the decrease in λ_1 observed at high-gravity levels could have resulted from suppression of convection in the bulk liquid. It is noted, however, that the change in spacing could not have resulted solely from a uniform high-gravity condition, and one might expect that Coriolis effects contribute to the observed spacing changes. Based on experimental conditions, a first approximation of the Coriolis acceleration, a_c, is calculated[17] to be $a_c = 0.02$ g. This calculated value is small compared to the acceleration at the solid/liquid interface (15.3 g) and how it might influence the axial flow is not obvious.

Altered Mass Transport

Good agreement between the measured primary dendrite arm spacing and theory resulted from changing the diffusion coefficient. Unfortunately, the actual fluid flow, and any possible alteration to it, was inferred and not directly observed. An alternative explanation for the observed decrease in primary dendrite arm spacing considers buoyancy of the rejected tin-rich liquid.

Referring to Fig. 5, tin is being rejected from a solidifying lead-rich dendrite. This solute-rich liquid is less dense than the bulk liquid. A simple analysis[17] showed that this liquid would rise, at one g, at a rate comparable to the dendrite growth velocity. As may be appreciated from Stokes's law:

$$v = \frac{2 \ g \ r^2(\rho_1 - \rho_2)}{9\eta} \tag{1}$$

the rise velocity, v, increases as g increases. Consequently, enhancing the gravity field is likely to extend the solute profile further ahead and less laterally, as shown in Fig. 6. This constrains growth of the secondary arms and promotes a deeper interdendritic pocket of the bulk liquid, C_0. This interdendritic liquid becomes increasingly undercooled and either nucleates solid, Fig. 7, or promotes growth of a tertiary branch, Fig. 8. Either mechanism will quickly establish a viable primary arm and reduce the spacing, Fig. 9. Evidence for this is seen in the microstructures of Fig. 10 and 11. The former (1 g) shows more widely spaced and intricate dendrites when compared to the latter (15.3 g).

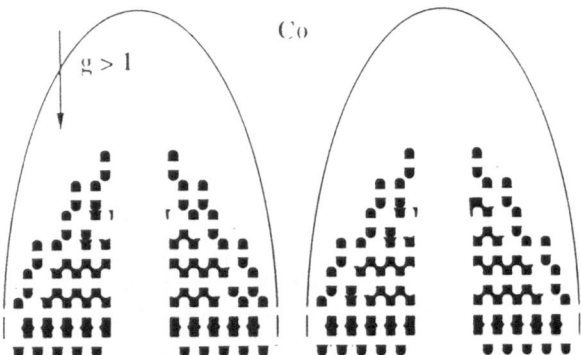

Figure 6. Same as Fig. 5, but with acceleration greater than 1 g.

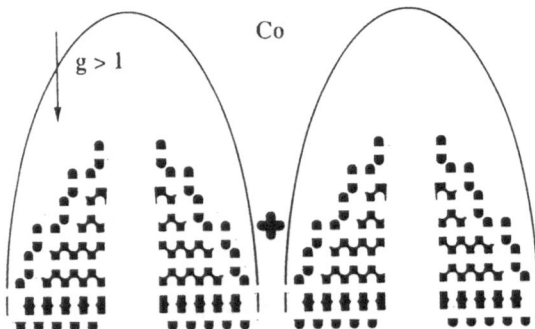

Figure 7. Schematic representation of solid nucleating in the undercooled interdendritic liquid.

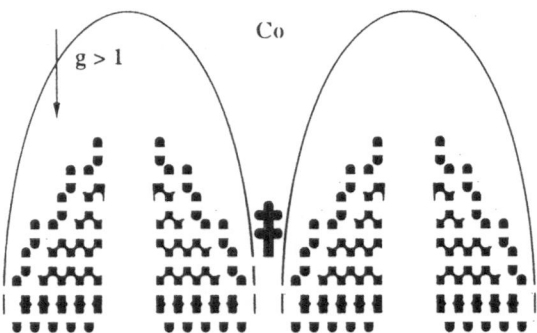

Figure 8. Schematic representation of a tertiary dendrite arm establishing itself as a primary.

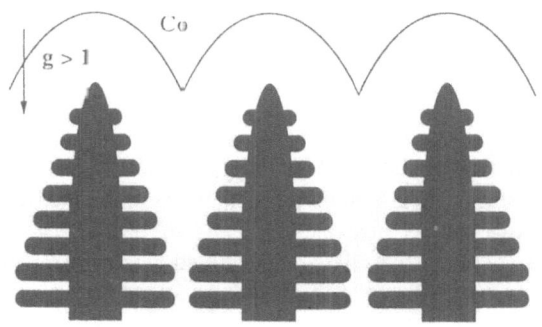

Figure 9. Schematic representation of steady-state dendrite growth with acceleration greater than 1 g.

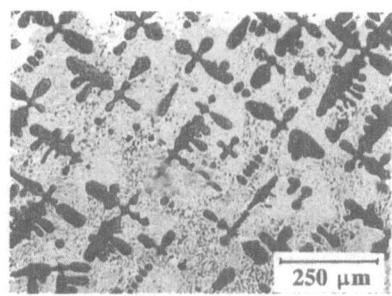

Figure 10. Cross-sectional micrograph of a Pb- 50 wt pct Sn alloy solidified in unit-gravity, growth velocity =21.1 $\mu.ms^{-1}$, temperature gradient =3.5 $K.mm^{-1}$.

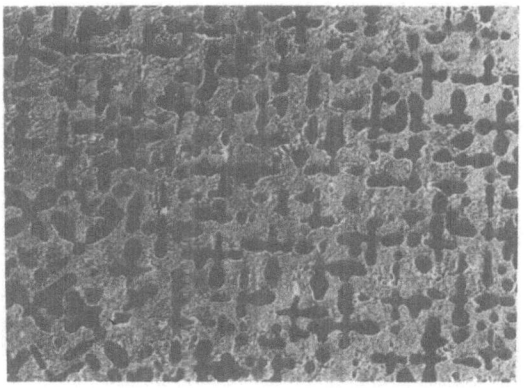

Figure 11. Same as Figure 10 except the imposed gravity level is 15.3 g.

It is relevant to note that current theory[21,24-26] links the equilibrium primary dendrite spacing to processing parameters as follows:

$$\lambda_1 = KG^{-1/2} V^{-1/4} C_o^{1/4} \tag{2}$$

Here G is the temperature gradient, V is the growth velocity, C_o is the composition, and K is a constant consisting of phase diagram parameters and material properties. Fig. 4 shows that a range of spacings is possible *independent* of the parameters given in Eq. (2). Consequently, convection effects likely contributed to the discrepancy found between theory and experiment.[27]

FUTURE STUDIES

As seen in Fig. 4, the primary dendrite arm spacing decreased as the imposed gravity level was increased. We attributed this to modification of convection in the bulk liquid, effectively suppressing solute diffusion and/or enhanced buoyancy of the rejected, solute-rich liquid. While these arguments appear reasonable, they are, in principle, based on inference, i.e. the presumed alterations in the liquid have not been observed. To this end an assembly was constructed that consists of an ampoule and optics that, with the aid of a laser, illuminates a cross-section of the contained fluid, as shown in Fig. 12.

The cell is filled with a transparent fluid, e.g. 10 cst silicone oil, and laced with tracer particles. A temperature gradient is imposed that induces a single convection cell. Flow, and changes in flow, are directly observed and recorded *during centrifugation*. An example of particle imaging in the cell section is seen in Fig. 13. Preliminary results indicate that convection flow patterns can be profoundly modified under the influence of the acceleration imposed by the centrifuge.

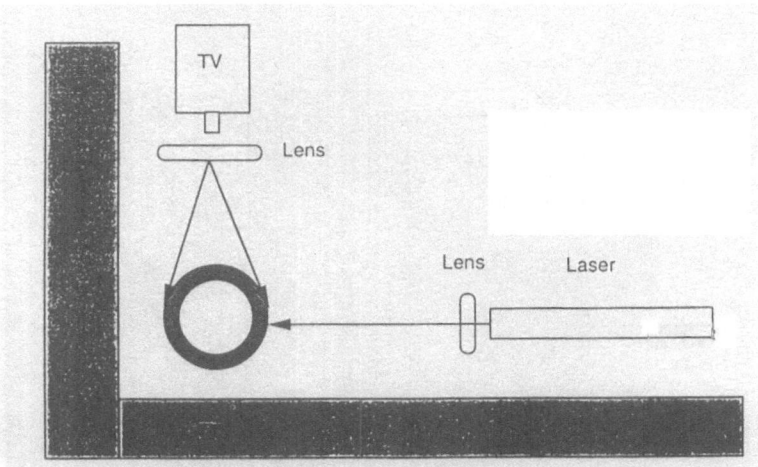

Figure 12. Schematic drawing of the transparent cell assembly.

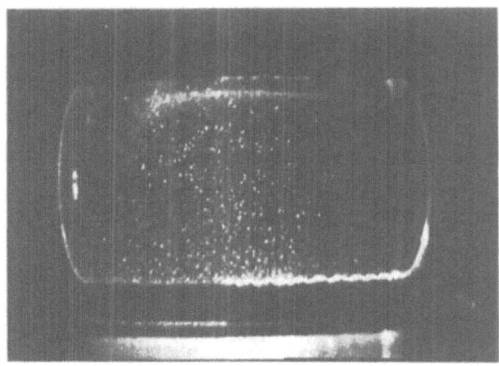

Figure 13. Particle imaging in the fluid cell.

CONCLUSIONS

1. At constant V and G, no perceptible changes in λ_2, λ_E, or d were found in directionally solidified lead- 50 wt pct tin alloys as a function of increasing gravity level.

2. The primary dendrite arm spacing decreased as a function of the increased gravity level imposed by centrifugation. This was attributed to 1) modification of convection in the bulk liquid that effectively suppressed solute diffusion and/or 2) enhanced buoyancy of the rejected, solute-rich liquid.

3. Centrifugation is capable of influencing a wide range of uniform dendritic spacings.

Acknowledgements

This work was sponsored by the NASA Office of Space Science and Applications and by NASA's JOVE project under Grant #NAG8-151. The technical assistance of Mr. Chris Apple is greatly appreciated.

REFERENCES

1. P.J. Shlichta and R.E. Knox, *J. Cryst. Growth* 3-4:808 (1968).
2. G. Müller, E. Schmidt and P. Kyr, *J. Cryst. Growth* 49:387 (1980).
3. G. Müller and G. Neuman, *J. Cryst. Growth* 59:548 (1982).
4. G. Müller and G. Neuman, *J. Cryst. Growth* 63:58 (1983).
5. W.E. Langlois, *J. Cryst. Growth* 63:67 (1983).
6. G. Müller, G. Neuman, and W. Weber, *J. Cryst. Growth* 70:78 (1984).
7. G. Müller, *J. Cryst. Growth* 99:1242 (1990).
8. W. Weber, G. Neuman and G. Müller, *J. Cryst. Growth* 100:145 (1990).
9. H. Rodot, L.L. Regel and A.M. Turtchaninov, *J. Cryst. Growth* 104:280 (1990).
10. B.T. Murray, S.R. Coriell and G.B. McFadden, *J. Cryst. Growth* 110:713 (1991).
11. L.L. Regel, M. Rodot and W.R. Wilcox, eds., "Material Processing in High Gravity," (1992).
12. M. McLean. "Directionally Solidified Materials for high Temperature Service," The Metals Society (1983).
13. J.D. Verhoeven, J.T. Mason and R. Trivedi, *Metall. Transactions A.* 17A:991 (1986).

14. C. Battaile, Masters Thesis, Vanderbilt University, Dept. of Materials Science and Engineering (1992).

15. C. Battaile, R.N. Grugel, A.B. Hmelo and T.G. Wang *in:* "Microstructural Design by Solidification Processing," E. J. Lavernia and N.N. Gungor, eds., The Minerals, Metals and Materials Society (1992) pp 161-172.

16. R.N. Grugel, *Materials Characterization* 28:213 (1992).

17. C. Battaile, R.N. Grugel, A.B. Hmelo and T.G. Wang: Submitted to *Metallurgical Trans. A.*

18. D.J. Allen and J.D. Hunt, *Metall.Transactions A.* 10A:1389 (1979).

19. T. Okamoto, K. Kishitake and I. Besso, *J. Cryst. Growth* 29:131 (1975).

20. T. Huang, D. Lu and Y. Zhou, *Acta. Astro* 17:997 (1988).

21. J.D. Hunt, "Solidification and Casting of Metals," Book 192, The Metals Society, London (1979) pp 3-9.

22. J. Liu, *Scripta Met.* 26:179 (1992).

23. G. Frohberg, K.H. Kraatz and H. Wever *in:* "Scientific Results of the German Spacelab Mission D1," P.R.Sahm, R. Jensen and M.H. Keller, eds., WPF, (1986) pp 144-151.

24. W. Kurz and D.J. Fisher, *Acta Metall.* 29:11 (1981).

25. R. Trivedi, *Metall. Transactions A.* 15A:977 (1984).

26. V. Laxmanan, *J. Cryst. Growth* 83:391 (1987).

27. W. Kurz and R.N. Grugel. "Materials Science Forum," Trans Tech. Publications, Switzerland (1991) pp 185-204.

THE ROLE OF THERMAL STRESS IN VERTICAL BRIDGMAN GROWTH OF CdZnTe CRYSTALS

Taipao Lee[1], John C. Moosbrugger[1], Frederick M. Carlson[1]
and David J. Larson, Jr.[2]

[1] Department of Mechanical and Aeronautical Engineering
Clarkson University
Potsdam, NY 13699-5729
[2] Grumman Corporate Research Center
Bethpage, NY 11714-3580

ABSTRACT

Computational studies of thermal fields and resulting thermoelastic stress fields were undertaken for the vertical Bridgman-Stockbarger growth of CdZnTe crystals. Companion experimental studies included the growth of crystals grown with the same process parameters and the same geometry as the process modeled in the computations. Characteristics of the crystals grown were compared with the computational predictions. Predictions of growth ampoule outer wall temperatures agree well with thermocouple data taken during the growth experiment. Additionally, the computed excess stress distribution resulting from the thermoelastic stress history in the solid is seen to agree qualitatively with synchrotron contour topography on a slice taken from the grown ingot. The computational models are shown to provide a good tool for the study of the influence of process parameters on the quality of crystals grown by this method, at least as far as thermal stress influences the defect distribution. The influence of low-g and high-g environments on growth is discussed.

INTRODUCTION

In the vertical Bridgman growth of semiconductor crystals, one of the factors controlling final crystal quality is the thermomechanical history imposed on the solidified material. This thermomechanical history arises because of thermal stresses due to nonuniform temperature fields, thermal expansion coefficient mismatch between the cooling solid crystal and the growth vessel if the crystal sticks to the container, and stresses due to the forces imposed by the combined weight of the melt and solid. The thermal contributions to stress can be controlled, to some extent, by careful control of the applied thermal history. However, to effectively control these factors, an understanding of the effects of the applied thermal history in the development of stresses for

Materials Processing in High Gravity, Edited by L.L. Regel
and W.R. Wilcox, Plenum Press, New York, 1994

the particular process is needed. The results of computational models of thermal fields and resulting thermal stresses can be used for this purpose. These models can follow a heirarchy ranging from simple to complex, depending on the realism of the assumptions involved in constructing them. Often, many simplifying assumptions must be made because there is little or no experimental information (property data, boundary conditions, etc.) forming a basis for more complex or realistic simulations. Process models, regardless of their level of sophistication, can be validated by process monitoring such as thermocouple measurements and by characterization of crystals grown using the same growth parameters.

In this paper, results of computational models and experimental results for the vertical Bridgman growth of CdZnTe crystals will be reported. The most important application for CdZnTe crystals is as a substrate for epitaxial growth of HgCdTe infrared detector material. For these applications, high quality (low defect density) single crystals are required. Attention will be focussed on the role of thermal stress in the growth of this material. Qualitative comparison of strain maps obtained using x-ray synchrotron diffraction contour topography will be made with a computed measure of the thermoelastic stress history.

EXPERIMENTS

Figure 1 shows a schematic of the furnace used for the directional solidification experiment which will be discussed. The material solidified was CdZnTe and the vertical Bridgman-Stockbarger technique was employed. A total of 24 control thermocouples was used for control of the power to the heaters enabling imposition of the outer wall temperature profile. Additionally, six thermocouples (independent of the furnace control system) measured the temperature of the outside of the quartz ampoule wall at the locations shown. Numerous translational sensors recorded the ampoule position for precise feedback control of the ampoule position. At the bottom of the ampoule, a 6 mm right circular cylinder seed crystal was inserted in the tapered well with an orientation of [111], corresponding to the growth axis. After initial partial meltback into the seed crystal, the entire ampoule containing the liquid charge was translated through the controlled temperature zone, effecting directional solidification.

Included in the characterization of the 38 mm ingots resulting from this growth process were x-ray double crystal rocking curve, x-ray precision lattice parameter, energy dispersive x-ray analysis, etch pit density, photoreflectance and synchrotron contour topography. In this paper, especially emphasized are synchrotron diffraction contour topography maps made on slices taken from the ingots. These synchrotron maps provide for a measure of the residual strain in the slice.[1]

MODELING

The heat transfer model uses the finite element method and is premised on the assumption that only conduction heat transfer occurs in the solid crystal, liquid melt and growth ampoule, with optical transmission (radiation) through the ampoule wall and across the gap to the outer wall. Fluid convection does not affect heat transfer significantly in this system because the Prandtl and Rayleigh numbers are small. Constant properties are assumed for the solid, melt, quartz ampoule, etc. The thermal properties used are listed in Tables 1 and 2. Applied boundary conditions are the temperatures of the control thermocouples (with linear interpolation between the thermocouple positions) along the outer wall with adiabatic conditions applied at the two ends. The furnace domain is chosen long enough so that the boundary conditions on the two ends

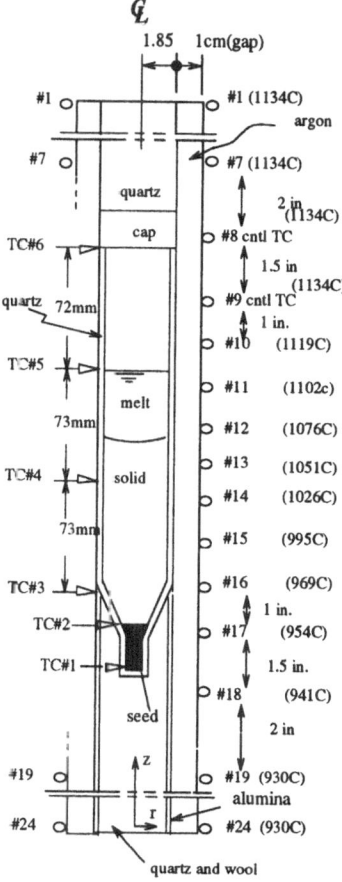

Figure 1. Schematic of the furnace and growth ampoule.

Table 1. System Thermal Conductivity :W/K m.

Part	Material	Temperature	Conductivity	Ref.
holder	Fiberfrax *	1300 K	0.3	
ampoule	quartz	300 K	1.5	[9]
support (outside)	alumina **	1300 K	6.0	
support (inside)	quartz and wool	300 K	0.04	[11]
argon	——	1366 K	0.052	[11]

⋆ Sohio Engineered Material Company
⋆⋆ WYLE Laboratories

will not affect the temperature field within the ampoule. The approximate thermal history is obtained by assuming steady state conditions for various ampoule locations within the controlled temperature zone. That is, a steady solution is obtained for a given ampoule location (for a specified outer wall temperature profile) and then the ampoule and charge are translated along the growth axis direction relative to the control thermocouple locations (by changing the outer wall temperature profile). This is repeated for incremental translations of the ampoule and charge, simulating the

Table 2. CdZnTe Properties.

Property	Units	Temperature	Values	Ref.
Melting Temperature T_m	K	—	1368	[9]
Thermal	W/K m	$T < 1368$ K (solid)	1.0	[10]
Conductivity		$T > 1368$ K (liquid)	2.0	[10,12]
C_{11}	MPa	—	53500	[5-8]
C_{12}	MPa	—	36900	[5-8]
C_{44}	MPa	—	20200	[5-8]
Thermal Expansion Coeff.	K^{-1}	—	0.0000065	[9]
CRSS	MPa	$800K < T < 1353K$	$23.07e^{-0.00337T(K)}$	[4]

continuous ampoule translation of the experiment. This is justified since the furnace temperature gradient and the growth rate were low (approximately 10 K/cm and 1.6 mm/h). Thus, we solve:

$$\nabla \cdot k \nabla T = 0 \tag{1}$$

where T is the temperature and k is thermal conductivity. Continuity of heat flux across material boundaries is also specified. For example, across the liquid-solid interface

$$k_s \nabla T \cdot \vec{n} = k_l \nabla T \cdot \vec{n} \tag{2}$$

where k_s and k_l are the thermal conductivities of the solid and liquid, respectively, and $\vec{n}$ is the unit normal to the solid-liquid interface. Note that latent heat is neglected since the growth rate is low. In the finite element method used, conductivity is constant within an element so that heat flux continuity across element boundaries is easily satisfied during the assembly of the system equations. Radiation across the 1 cm argon gap is taken to be one-dimensional (since the gap aspect ratio is large) and is taken into account using the equivalent conductivity technique. Emissivities used are 0.5 and 0.4 for the quartz and furnace wall, respectively.

The interface shape and location for any steady solution is found by iteration as follows: First, the position of points along the interface are guessed. Then, the thermal conductivity is assigned for elements on either side of the interface. The steady solution is obtained and the location of the computed 1368 K isotherm, corresponding to the melting temperature, is compared to the guessed location. The thermal conductivities of the elements are then changed until the computed and guessed interface locations coincide.

Figure 2 shows a schematic of the applied boundary conditions and solution domain for a given ampoule location. Note that axisymmetry is used to obtain the steady temperature field. For the thermo-elastic stress field, a fully three-dimensional finite element model is used, since the material elasticity properties are anisotropic.

For the computation of the thermo-elastic stress field, we solve equilibrium:

$$\nabla \cdot \boldsymbol{\sigma} = 0 \tag{3}$$

along with the strain-displacement relations:

$$\epsilon = \frac{1}{2}(\nabla \vec{u} + \nabla \vec{u}^T) \tag{4}$$

and the constitutive equations:

$$\boldsymbol{\sigma} = \mathbf{C} : (\epsilon - \alpha \Delta T \boldsymbol{\delta}). \tag{5}$$

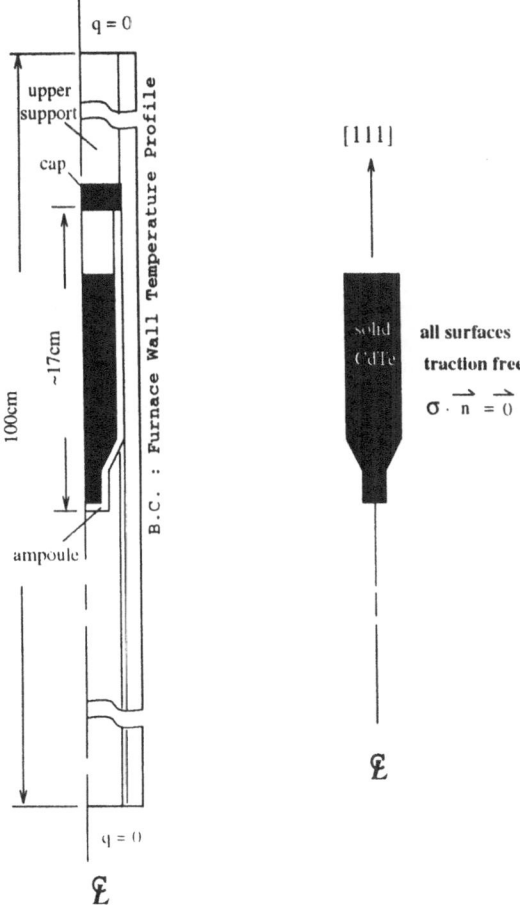

Figure 2. Thermal and stress boundary conditions.

In Eqs. (3-5), σ is the stress tensor, ϵ is the infinitesimal strain tensor, $\vec{u}$ is the displacement field, superscript T denotes the transpose, $\mathbf{C}$ is the tensor of elastic moduli, α is the coefficient of thermal expansion, $\Delta T = T - T_m$, T_m is the melting temperature, and δ is the identity tensor. Since CdZnTe crystals possess cubic symmetry, in matrix (Voight) notation, Eq. (5) can be written as:

$$
\begin{bmatrix} \sigma_{11} \\ \sigma_{22} \\ \sigma_{33} \\ \sigma_{23} \\ \sigma_{13} \\ \sigma_{12} \end{bmatrix} = \begin{bmatrix} C_{11} & C_{12} & C_{12} & 0 & 0 & 0 \\ C_{12} & C_{11} & C_{12} & 0 & 0 & 0 \\ C_{12} & C_{12} & C_{11} & 0 & 0 & 0 \\ 0 & 0 & 0 & C_{44} & 0 & 0 \\ 0 & 0 & 0 & 0 & C_{44} & 0 \\ 0 & 0 & 0 & 0 & 0 & C_{44} \end{bmatrix} \begin{bmatrix} \epsilon_{11} - \alpha\Delta T \\ \epsilon_{22} - \alpha\Delta T \\ \epsilon_{33} - \alpha\Delta T \\ \gamma_{23} \\ \gamma_{13} \\ \gamma_{12} \end{bmatrix}
\tag{6}
$$

where the stress and strain components correspond to a [100] crystallographic axis aligned, material Cartesian coordinate system. Note that in Eq. (6), engineering shear strain $\gamma_{ij} = 2\epsilon_{ij}$ for $i \neq j$ is used. This is necessary to preserve symmetry in the matrix of elastic moduli. Values of the thermal expansion coefficient and elastic moduli used are given in Table 2. The elasticity tensor (or matrix) components are transformed[2,3] for the global coordinate system used in the computations, in which the z-axis corresponds to a [111] crystallographic direction.

For each steady temperature field, temperatures are mapped from the axisymmetric solution to the three dimensional grid used for the elasticity solution. This temperature field is used to solve for the thermal stresses associated with the solution to Eqs. (3)-(6) along with the boundary conditions illustrated in Figure 2. The resulting thermal stress is used to compute the so-called "excess stress," which is the maximum of the resolved shear stress over all of the twelve {111} ⟨110⟩ slip systems given in Figure 3, i.e.:

$$\tau_{xs} = max[\vec{\nu}^j \cdot \boldsymbol{\sigma} \cdot \vec{n}^j] - CRSS \tag{7}$$

where $\vec{\nu}^j$ and $\vec{n}^j$ are unit vectors in the slip direction and slip plane normal directions for the jth slip system, respectively. The CRSS is the critical resolved shear stress, for which the temperature correlation given in Table 2 is based on the data of Balasubramanian and Wilcox.[4] This quantity is the excess stress τ_{xs} for the element (stress components are constant within an element for second order elements) for the previously computed temperature field.

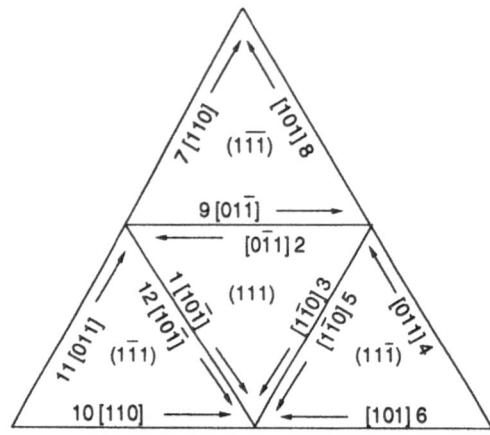

Figure 3. Twelve {111} ⟨110⟩ slip systems used for computation of τ_{xs} shown on the Thompson tetrahedron.

RESULTS AND DISCUSSION

Figures 4 and 5 show the computed temperatures at the six thermocouple locations. These are shown superimposed on the measured values and are plotted versus time. Note that the computed temperatures correspond to steady solutions for given ampoule locations or process times. As can be seen in Figures 4 and 5, the computed temperatures at these locations are within 5 K of the measured values over approximately 70 h simulated. Figure 6 shows the maximum, over ten ampoule positions simulating the complete solidification history, of the excess stress computed for various slices through the crystal domain. Figure 7 shows the distribution of the maximum excess stress over the history for a cross section perpendicular to the growth direction and for a plane parallel to and passing through the cylindrical boule axis. Note that the highest levels of these maximum excess stresses occur at the periphery and center of the boule, corresponding to typical observations of dislocation density distributions in crystals grown by the vertical Bridgman method. Etch pit studies showed that these regions contain numerous subgrains. Also, higher levels of maximum excess stress occur

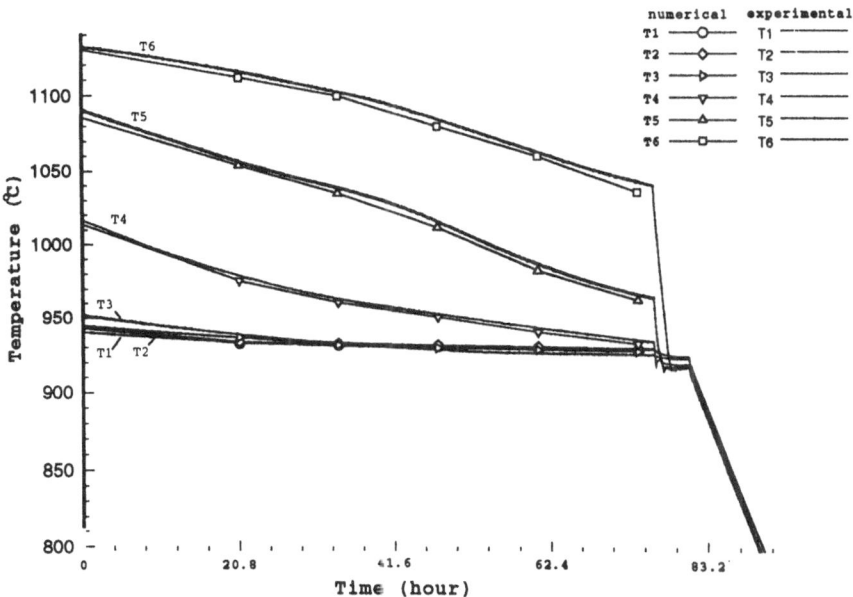

Figure 4. Computed temperature superimposed on measured temperature for the six thermocouple locations shown in Figure 1.

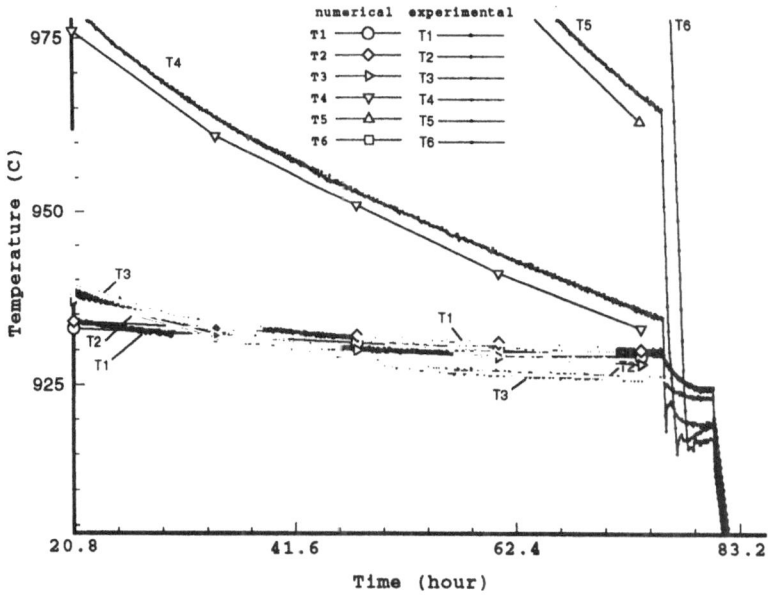

Figure 5. Computed temperature superimposed on measured temperature for the six thermocouple locations shown in Figure 1.

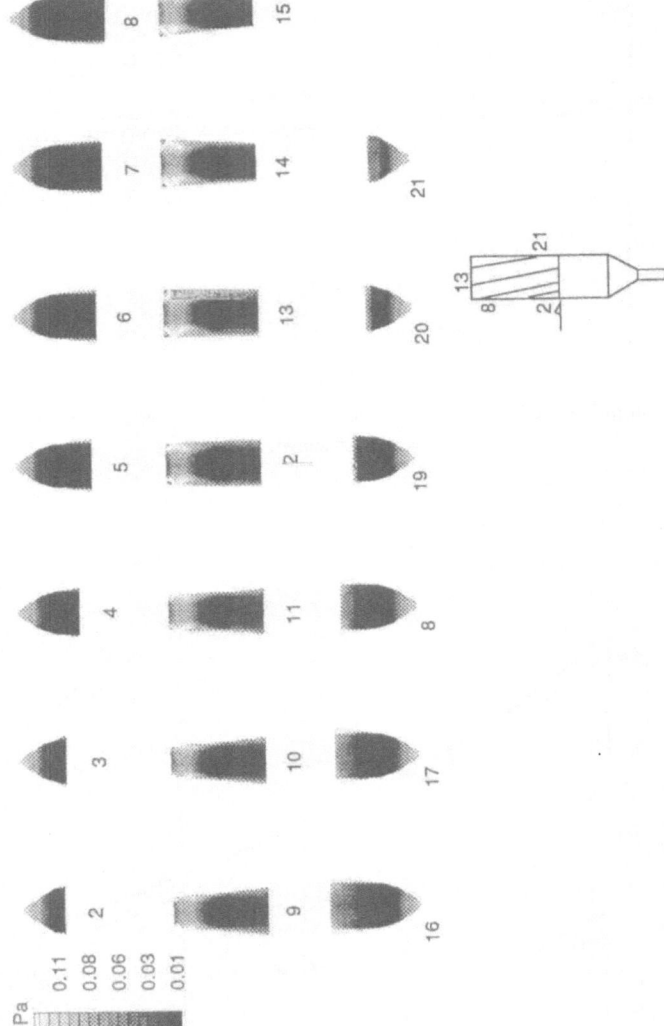

Figure 6. Maximum excess stress over the solidification history for various slices through the ingot.

118

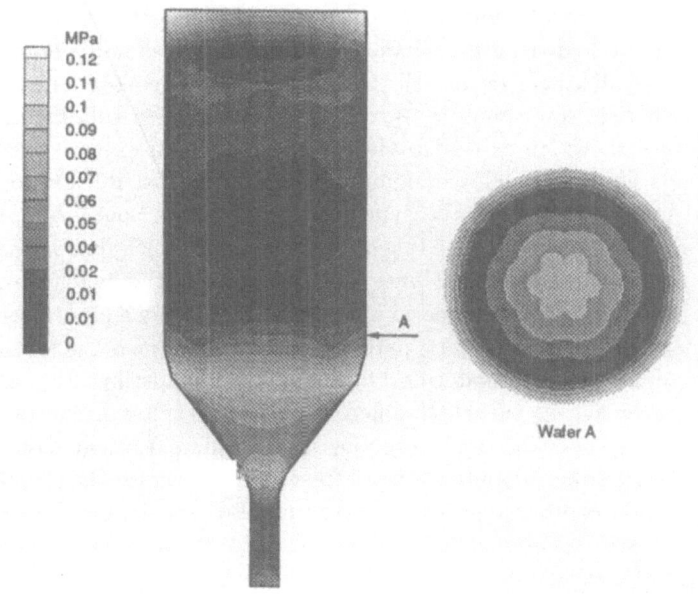

Water A

Figure 7. Excess stress distribution for cross-sections perpendicular and parallel to the growth axis.

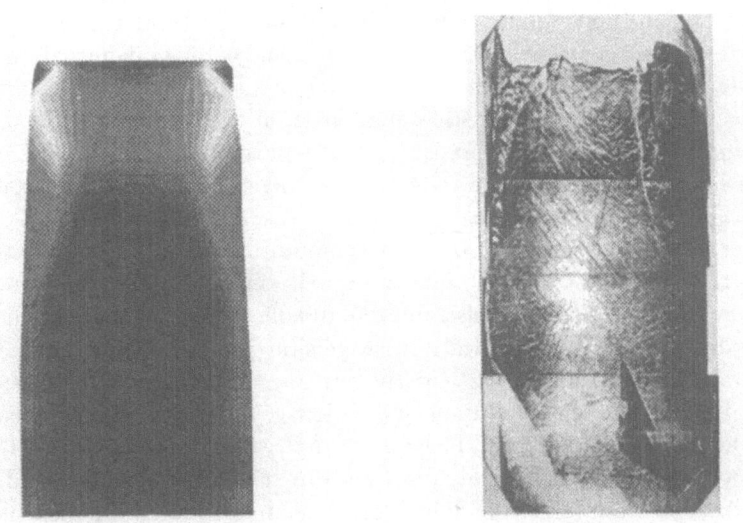

Excess Stress for Slice 12   Syncrotron Strain Map for Slice 12

Figure 8. Maximum Excess Stress Distribution and Synchrotron Diffraction Contour Topography Map for Slice 12.

near the tapered end (or "shoulder") of the ampoule and near the terminal transient (i.e. near the top free surface of the boule where the advancing solid-liquid interface has exhausted the melt). The high level of maximum excess stresses generated near the shoulder and terminal transient, with low stresses in between, indicates that ingots with relatively high length-to-diameter aspect ratios should be grown with this technique since small aspect ratios will yield ingots that have experienced high thermal stresses everywhere at some point in the solidification and post-solidification cool-down history. Figure 8 shows an expanded view of the maximum excess stress distribution for slice 12 (see Figure 6). This maximum excess stress distribution for slice 12 can be compared qualitatively with the synchrotron diffraction contour topography map made on an identically located slice taken from the ingot grown during the experiment. The section containing the hashed lines in the middle of the synchrotron diffraction contour topography map of the slice contains relatively low defect, low strain material. The sections exhibiting the spaghetti patterns on both sides near the terminal transient contain more highly strained material. On both sides, the distinct lines are boundaries between sections which have slightly different crystallographic orientations within the slice. The clear area near the top is a section with a different orientation that does not image. In comparing the computed excess stress distribution for the slice 12 to the synchrotron diffraction contour topography map, qualitative agreement is apparent. The computed excess stresses are highest in the same regions where the slice has the highest strains. Also the distinct lines or boundaries between sections with slightly different crystal orientations correspond well with the maximum excess stress patterns. It is noted that thermoelastic models cannot predict residual strains as are mapped by the synchrotron monochromated topography. Residual strains, in fact, are generated precisely because *inelastic* deformation has taken place and are the result of the complete temperature-deformation history in the solid. However, thermoelastic stresses, due to thermal gradients and/or thermal expansion mismatch between the growth ampoule and an adhering solid undergoing cooling, drive the inelastic deformation. Therefore, computed thermoelastic stress fields can be a good indicator of deformation patterns. They provide us with a good engineering tool for predicting process parameter influence on defects influenced by inelastic deformation in the solid state. Also, they are relatively simple to perform in comparison to inelastic analyses, which require accurate constitutive models over the entire temperature range of interest and a concomitant orders-of-magnitude increase in computational effort.

Not reported in this paper are ongoing computational studies of the influence of adhesion of the solid crystal to the quartz growth ampoule under the same thermal history. These indicate that the constraint due to adhesion raises the maximum excess stress levels by an order-of-magnitude, with general patterns being similar to those reported here. These results are particularly germane to the question of crystal quality for growth in low-g or high-g environments. In low-g, the melt will have less tendency to contact the ampoule surface and hence stick to it when it has solidified. Conversely in high-g, such as growth under centrifugation, this tendency to contact and stick will be magnified, due to the large centrifugal forces tending to force the melt against the container walls. Thus, ampoule coating strategies intended to ameliorate the effect of sticking to the growth container will be more important for growth in a high-g environment. Preliminary characterizations of crystals grown in low-g during the USML-1 Space Shuttle mission, on the other hand, have yielded very low defect density material near regions in which the solid did not contact the ampoule.

CONCLUSIONS

It can be concluded that the quasisteady thermal and thermoelastic stress models provide relatively simple predictive tools to study the influence of processing variables on crystal quality as influenced by thermal stress for the growth of CdZnTe grown by the vertical Bridgman-Stockbarger technique. It can be concluded that ingots grown with this technique should be of relatively large aspect ratio. Any process variables that increase the tendency for the melt to contact the growth ampoule and stick to it when solidified will be detrimental to crystal quality, since constraint due to sticking increases the magnitude of the maximum elastic excess stresses by an order-of-magnitude.

Acknowledgements

The support of this project by the NASA Microgravity Science and Applications Division (contract no. NAS8-38147) is gratefully acknowledged. The authors also wish to acknowledge the contributions of G. Long, D. Black and B. Steiner of the National Institute of Standards and Technology for the x-ray synchrotron contour topography.

REFERENCES

1. D.J. Larson, Jr., R.P. Silberstein, D. DiMarzio, F.M. Carlson, D. Gillies, G. Long, M. Dudley and J. Wu, Compositional strain contour and property mapping of CdZnTe boules and wafers, *Semicond. Sci. Technol.* 8:911 (1993).

2. S. Miyazaki, Elastic constant matrix required for thermal stress analysis and bulk single crystals during Czochralski growth, *J. Crystal Growth* 106:149(1990).

3. R.F.S. Hearmon, "An Introduction to Applied Anisotropic Elasticity," Oxford University Press, London. (1961)

4. R. Balasubramanian and W.R. Wilcox, Mechanical properties of CdTe, *Mat. Sci. Eng.* B16:1 (1993).

5. H.J. McSkimin and D.G. Thomas, Elastic moduli of cadmium telluride, *J. Appl. Phys.* 33:56 (1962).

6. D. Berlincort, H. Jaffe and L.R. Shiozawa, Electroelastic properties of the sulfides, selenides, and tellurides of zinc and cadmium, *Phys. Rev.* 129:1009 (1963).

7. Yu.Kh. Vekilov and A.P. Rusakov, Elastic constants and lattice dynamics of some $A^{II}B^{VI}$ compounds, *Sov. Phys. - Solid State* 13:956 (1971).

8. R.D. Greenough and S.B. Palmer, The elastic constants and thermal expansion of single-crystal CdTe, *J. Phys. D: Appl. Phys.* 6:587 (1973).

9. C. Parfeniuk, F. Weinberg, I.V. Samarasekera, C. Schvezov and L. Li, Measured critical resolved shear stress and calculated temperature and stress fields during growth of CdZnTe, *J. Crystal Growth* 119:261(1992).

10. T. Jasinski, and A.F. Witt, On control of the crystal-melt interface shape during growth in a vertical Bridgman configuration, *J. Crystal Growth* 71:295(1985).

11. J. H. Lienhard, "A Heat Transfer Textbook," Prentice-Hall, Englewood Cliffs, New Jersey (1981)

12. P. Rudolph and Manfred Muhlberg, Basic problems of vertical Bridgman growth of CdTe, *Mat. Sci. Eng.* B16:8(1993).

MORPHOLOGICAL STABILITY OF DIRECTIONAL SOLIDIFICATION IN A CENTRIFUGAL FIELD

Valentin S. Yuferev

A.F. Ioffe Physico-Technical Institute
194021 St. Petersburg, Russia

ABSTRACT

Morphological stability of crystallization front is considered in the case when solutal convection in a melt is driven by the centrifugal and Coriolis forces. Temperature field is assumed to be a linear function and a vector of an angular rate is perpendicular to a growth direction. It is shown that the effect of Coriolis force has a dual character. At first, increase of Coriolis force lead to an instability of crystallization front. But when Taylor number exceeds some critical value, crystallization front becomes stable again and Coriolis force surpass practically completely the influence of the convection.

INTRODUCTION

When simulating a high gravity in a centrifuge we are faced with the Coriolis force, which is usually absent in conventional material processing. The Coriolis force influences flows in a fluid, and through the convection influences the quality of grown crystals. The influence of the Coriolis force on convection in a melt has been considered.[1-5] It has been shown[1,2] that the Coriolis force can lead to an abrupt change of flow pattern. In particular, the convection was stabilized when the Taylor number exceeded a critical value. This phenomenon seems to be very important, and it is natural to ask whether the Coriolis force always stabilizes melt motion or whether its influence has a more complicated character. To answer this question it is necessary to investigate the stability problem.

There are many papers concerned with the stability of convection in directional solidification, e.g., Eq. (6a). But the same problem including the Coriolis force has not been investigated so far. Therefore, in this article we treat a quite simple case of the morphological stability of a solidification front in the presence of solutally-driven buoyant convection.

FORMULATION OF THE MODEL

We consider the situation where the mean position of the interface (z'=0) moves at constant speed V in the positive direction of the z'-axis (Fig.1). For simplicity, the following assumptions are made:

1. The temperature field in both the melt and the solid is a linear function of z', $T = T_0 + G_L z'$.
2. The latent heat of fusion is neglected.
3. The density of crystal and melt are equal.
4. Acceleration due to earth's gravity is not taken into account.
5. The rotation axis is perpendicular to the growth direction.

As shown later, under these assumptions the base state of our model can be considered to be motionless. In all other cases this is not true. Notice also that, in our formulation, the Coriolis force lays in the Y'0'Z'-plane.

We used conventional scaling:

$$
\begin{pmatrix} x' \\ y' \\ z' \end{pmatrix} = \ell_D \begin{pmatrix} x \\ y \\ z \end{pmatrix} ; \quad h' = \ell_D h ; \quad \begin{pmatrix} u' \\ v' \\ w' \end{pmatrix} = V \begin{pmatrix} u \\ v \\ w \end{pmatrix} ;
$$

$$
t' = t \frac{\ell_D}{V} ; \quad c' = \frac{c_s}{k} + \frac{G_C D}{v} c ; \tag{1}
$$

$$
G_C = \frac{k-1}{k} \frac{V}{D} c_s ; \quad \ell_D = \frac{V}{D} ;
$$

where ℓ_D is the diffusion length and G_C is the gradient of the impurity concentration at the crystallization front in a basic unperturbed state.

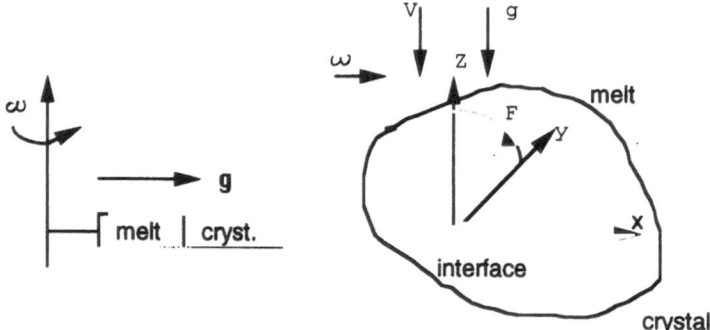

Figure 1. Schematic model of directional solidification in a centrifuge. Here g is the centrifugal acceleration, w is the angular rate, V is the growth rate and F is the Coriolis force.

In the Boussinessq approximation, the governing equations take the form:

$$\nabla \cdot u = 0 \tag{2a}$$

$$Sc^{-1}\left(u_t + (u \cdot \nabla) u - u_z\right) - \nabla u^2 + \nabla P +$$
$$+ R \cdot c \left(e_z - \varepsilon (e_y \times e_z)\right) + Q (e_y \times u) = 0 \tag{2b}$$

$$c_t + (u \cdot \nabla) - c_z - \nabla^2 c = 0 \tag{2c}$$

Here $Sc = v/D$ is the Schmidt number,

$$R = \frac{(1-k) C_s g \beta}{k v D} \left(\frac{D}{V}\right)^3 \tag{3a}$$

is the Rayleigh number, and

$$\varepsilon = \frac{2 \omega V}{g}$$
$$Q = \frac{2 \omega}{v} \frac{D^2}{V^2} = 2 Ek^{-1} = 2' Ta^{1/2} , \tag{3b}$$

e_y and e_z are the unit coordinate vectors and $\beta = - (1/\rho)(d\rho/dc)$ is the solutal expansion coefficient. If the solute is less dense than the solvent, β is positive. So $R > 0$ corresponds to an unstable density profile. On the other hand, $R < 0$ is for a stable one.

Parameter Q is usually quite large in centrifugal experiments. For example, for the typical values: $V = 10^{-4} \div 10^{-3}$ cm/s, $\omega = 1 \div 10$ s^{-1}, $D = 10^{-4}$ cm^2/s, and $Sc = 10$; Q lays in the range 20 to 2×10^4. Parameter ε is very small: $\varepsilon \approx 10^{-6}$ to 10^{-7}. Therefore, the corresponding term in Eq. (2b) can be neglected. That is the reason why the steady-state base solution of our equations does not include any fluid motion.

SOLUTION

Using standard manipulation for hydrodynamical stability analysis to remove the pressure, and presenting solution in the form:

$$\begin{pmatrix} u \\ v \\ w \\ c \end{pmatrix} = \begin{pmatrix} 0 \\ 0 \\ 0 \\ 1 - e^{-z} \end{pmatrix} + \begin{pmatrix} u(z) \\ v(z) \\ w(z) \\ c(z) \end{pmatrix} \exp\left(\sigma t + ia_x x + ia_y Y\right) , \tag{4}$$

where σ is the complex growth rate and a_x, a_y are the x- and y-components of a wave number, we obtain a set of equations for disturbances:

$$\mathcal{L}_1 c = w \exp(-z) \tag{5a}$$

$$\mathcal{L}_2 u + Q \left(ia_x \frac{dv}{dz} + a_x a_y w \right) = - ia_x R \frac{dc}{dz} \tag{5b}$$

$$\mathcal{L}_2 v + Q \left(ia_y \frac{dv}{dz} - a_x^2 w + \frac{d^2 w}{dz^2} \right) = - ia_y R \frac{dc}{dz} \tag{5c}$$

$$\mathcal{L}_2 w + Q \left(a^2 v - ia_y \frac{dw}{dz} \right) = - a R^2 c \tag{5d}$$

where $a^2 = a_x^2 + a_y^2$ and $\mathcal{L}_1$, $\mathcal{L}_2$ are the differential operators:

$$\mathcal{L}_1 = \frac{d}{dz^2} + \frac{d}{dz} - a^2 - \sigma \tag{6a}$$

$$\mathcal{L}_2 = \left(\frac{d}{dz^2} - a^2 \right) \left(\frac{d}{dz^2} - a^2 - Sc^{-1} \left(\sigma - \frac{d}{dz} \right) \right) \tag{6b}$$

The dimensionless boundary conditions for these equations are:

at infinity, $z \rightarrow \infty$:   u, v, w, c = 0

at the interface, $z = 0$: $w = v = u = \dfrac{dw}{dz} = 0$ $\tag{7a}$

$$\frac{dc}{dz} + (1 - k) c = \frac{\sigma + k}{M^{-1} - 1 + a^2 \Gamma} c \tag{7b}$$

where

$$\Gamma = \frac{T_m \gamma}{L m G_C} \left(\frac{v}{D} \right)^3$$

$$M = \frac{m |G_C|}{G_L} \tag{7c}$$

Equations (5b–d) are fourth–order. Therefore, in the general case, two additional boundary equations at the interface must be specified. For such conditions we used the following:

$$\text{at } z = 0 \qquad \frac{\partial}{\partial z} \left(\text{div} \, (u) \right) = 0$$

$$\text{rot}_z \left(L(u) \right) = 0 \tag{8a}$$

where $L(u)$ is the left side of the equation of momentum conservation [Eq. (2b)]. The conditions [Eq. (8a)] ensure that solution of Eqs. (5b–d) satisfies the equations:

$$\text{div} \, (u) = 0 \quad \text{and} \quad \text{rot} \left(L(u) \right) = 0 \tag{8b}$$

everywhere over the melt region.

On the other hand, Eqs. (5c–d) are connected between themselves but independent of Eq. (5b). So, to consider only Eqs. (5c and 5d) it is necessary to obtain a boundary condition independent of the velocity component u. For this purpose we combined Eqs. (3a and 8b) in the following way:

$$\text{rot}_z \left(L(u) \right) - \frac{a_y}{a_x} \left(\text{Sc}^{-1} \frac{\partial}{\partial z} \left(\text{div} \, (u) \right) + \frac{\partial^2}{\partial z^2} \left(\text{div} \, (u) \right) \right) = 0 \tag{9}$$

From here we shall have:

$$a^2 \frac{d}{dz} \left(\frac{dv}{dz} + \text{Sc}^{-1} v \right) = i a_y \frac{d^2}{dz^2} \left(\frac{dw}{dz} + \text{Sc}^{-1} w \right) \tag{10}$$

So, unlike the usual formulation for stability problems, when it is enough to consider only an equation for the z–component of velocity, w, in our case we have to solve two equations for the z and y components, w and v.

Solution of the problem was sought in the form:

$$\begin{pmatrix} c \\ w \\ v \end{pmatrix} = \sum_{1=0}^{\infty} \sum_{k=1}^{5} \begin{pmatrix} \alpha_{1k} \\ \beta_{1k} \\ \gamma_{1k} \end{pmatrix} \exp \left(- (r_k + 1) z \right) \tag{11}$$

Substituting the expansion Eq. (11) into Eq. (5) yields, firstly, two algebraic equations for determination of the coefficients r_k:

$$L_1(r_1) = r_1^2 - r_1 - a^2 - \sigma = 0 \tag{12a}$$

$$\det \begin{vmatrix} L_2(r_k) - ia_y r_k Q & Q\left(r_k^2 - a_x^2\right) \\ Q\, a^2 & L_2(r_k) + ia_y r_k Q \end{vmatrix} = 0 \tag{12b}$$

where $L_2(r_k) = (r_k^2 - a^2)(r_k^2 - a^2 - Sc^{-1}(\sigma + r_k))$ with $k = 2,3,4,5$, and secondly, relationships between the expansion coefficients α_{1k}, β_{1k}, γ_{1k}:

$$\left.\begin{aligned} \alpha_{ok} &= 0 \\ Q\, a^2 \gamma_{ok} &= -L_2(r_k) - i\, Q\, a_y r_k \end{aligned}\right\} \quad k = 2,3,4,5 \tag{13a}$$

$$\begin{aligned} a_{1+1,k} L_1\left(r_k + \ell + 1\right) &= \beta_{1,k}, \quad \ell = 0,1,\ldots \quad k = 1\ldots5 \\ \beta_{1k}\left(L_2(r_k + \ell) + iQa_y(r_k + \ell)\right) &+ a^2 R\, \alpha_{1k} + Q^2 a^2 \gamma_{1k} = 0 \end{aligned} \tag{13b}$$

$$\gamma_{1k}\left(L_2(r_k + \ell) - iQa_y(r_k + \ell)\right) - ia_y R(r_k + \ell)\alpha_{1K} -$$

$$- Q\left(a_x^2 - (r_k + \ell)^2\right)\beta_{1k} = 0, \quad \ell = \begin{cases} 0,1,\ldots & k = 1 \\ 1,2,\ldots & k = 2..5 \end{cases} \tag{13c}$$

After some transformations, Eq. (12b), we obtain the following equations for calculation of the roots r_k:

$$(r_k^2 - a^2)\left(r_k^2 - a^2 - Sc^{-1}(\sigma + r_k)\right)^2 = Q^2 a_x^2, \quad k = 2\ldots4 \tag{13d}$$

$$r_5 = a$$

Relationships Eq. (13) allow us to express all expansion coefficients α_{1k}, β_{1k}, γ_{1k} through the five coefficients α_{1k} ($k = 2,..5$). The latter four of these, α_{1k} ($k = 2,..5$), can be expressed through α with the help of boundary conditions Eqs. (7a) and (10). Applying the boundary condition for diffusion, Eq. (7b), then leads to an equation that relates the growth rate σ to other parameters. In the general case, $\sigma = \sigma_r + i\,\sigma_i$. So specifying $\sigma_r = 0$, we obtain a dispersion relation for the neutral stability of disturbances with wave number a.

RESULTS OF CALCULATIONS

We calculated the curves for neutral stability considering M and σ_i as functions of the other independent parameters. Parameters Γ and Sc were set equal to 0.1 and 10, respectively. The basic calculations were made for a segregation coefficient $k = 0.01$. Results of the calculations are presented in Figs. 2–4.

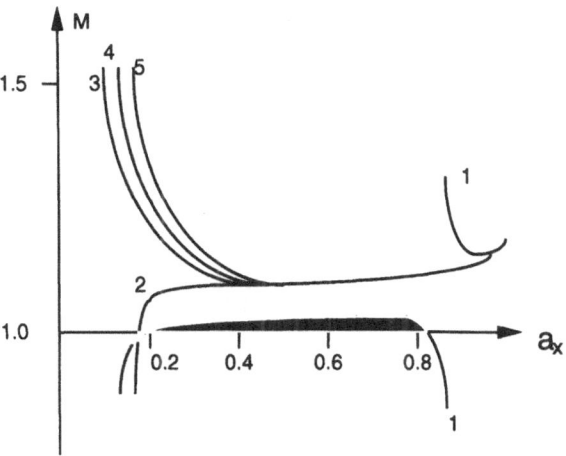

Figure 2. The stationary branches of the neutral stability curves for an unsteady density profile: R = 20, a_y = 0. (1) Q = 0; (2) Q = 5; (3) Q = 20; and (4) Q = 40. Curve (5) corresponds to R = 0.

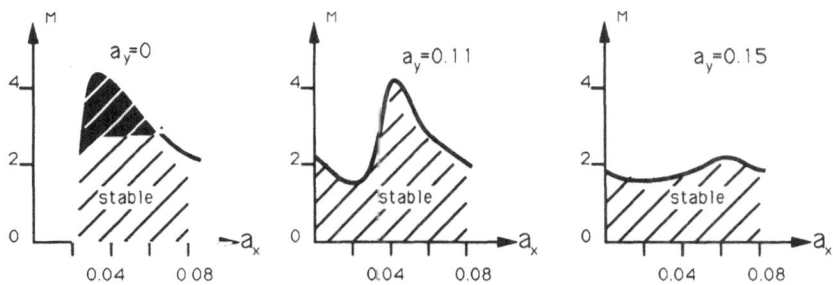

Figure 3. Neutral stability curves for R = 20, Q = 40, and different values of a_y (two-dimensional disturbances).

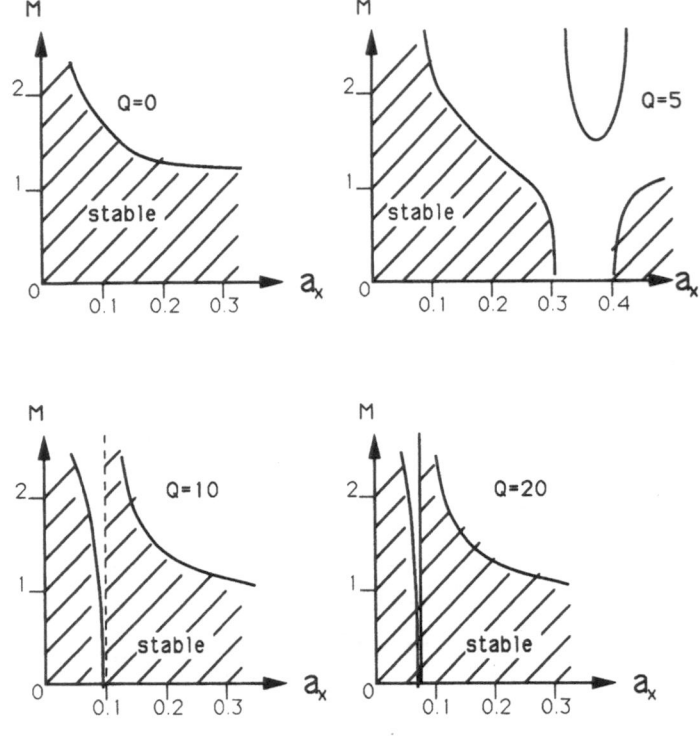

Figure 4. The neutral stability curves for a steady density profile: $R = -5$, $a_y = 0$.

Figure 2 shows the case when a density profile is unstable ($R>0$). In line with the results of Hurle,[7] the Rayleigh number lies above a critical value so that a plane solidification front is unstable too. In Fig. 3 this instability manifests itself as the neutral stability curve crossing the abscissa axis. In this case, a hydrodynamic instability of the melt takes place. Figure 2 shows, also, that with an increase of the Coriolis force, the neutral curves tend to the curve described by the Mullins–Sekerka theory when convection is absent ($R = 0$). Nevertheless, strictly speaking, even for large values of Q the crystallization front remains unstable for small wave numbers, as seen in Fig. 3. But the range of instability with increasing of Q is displaced in the direction of zero wave number, so that in practice, for an interface of finite size, the interphase boundary becomes stable. In this figure, the neutral stability curves for two–dimensional disturbances, when a_y is not equal zero, are shown also. Figure 4 corresponds to $R = -5$ when the density profile is stable. Here, at first, the action of the Coriolis force lead to instability of the crystallization front. But then, as Q increases, the stability of the front begins to rise again and the interval of hydrodynamically unstable wave numbers becomes narrower and is displaced in the direction of very small values. So we obtain again that the crystallization front of finite size becomes stable when Q exceeds some critical value.

CONCLUSION

The effect of Coriolis force on the morphological stability of directional solidification has a dual character. At intermediate values of the parameter Q (Taylor number) the Coriolis force leads to an instability of the crystallization front. But at large values of this parameter, the Coriolis force, on the contrary, stabilizes the interphase boundary and suppresses convection practically completely.

REFERENCES

1. W. Weber, G. Newmann, and G. Müller, Stabilizing influence of the Coriolis force during melt growth on centrifuge, *J. Cryst. Growth* 100:145 (1990).
2. G. Müller, G. Newmann, and W. Weber, The growth of homogeneous semiconductor crystals in a centrifuge by stabilizing influence of the Coriolis force, *J. Cryst. Growth* 119:12 (1992).
3. W.A. Arnold, W.R. Wilcox, and F. Carlson, A. Chait, and L.L. Regel, Transport modes during crystal growth in a centrifuge, *J. Cryst. Growth* 119:24 (1992).
4. M.A. Fikri, G. Labrosse, and M. Betroni, The melt phase hydrodynamics for the "stabilized" Bridgman procedure applied under centrifugation: preliminary analysis and numerical results, *J. Cryst. Growth* 119:41 (1992).
5. P.A. Vorobiev and N.A Baturin, Laminar convection in the melt during crystal growth in a centriguge, *J. Cryst. Growth* 119:111 (1992).
6. S.R.Coriell, M.R. Cordes, W.J. Boettinger, and R.F. Sekerka, Convective and interfacial instabilities during unidirectional solidification of a binary alloy, *J. Cryst. Growth* 49:13 (1980).
7. D.T.J. Hurle, E. Jakeman, and A.A. Wheller, Hydrodynamic stability of the melt during solidification of a binary alloy, *Phys. Fluids* 26:624 (1983).
8. G.B. McFadden, S.R. Coriell, R.F. Boisvert, M.E. Gliksman, and Q.T. Fang, Morphological stability in the presence of fluid flow in the melt, *Met. Trans. A.* 15:2117 (1984).
9. J.J. Favier and A. Rouzand, *J. Cryst. Growth* 367:64 (1983).

EFFECTS OF CORIOLIS AND CENTRIFUGAL FORCES ON THE MELT DURING DIRECTIONAL SOLIDIFICATION OF A BINARY ALLOY

D.N. Riahi

Department of Theoretical and Applied Mechanics
216 Talbot Laboratory
104 S. Wright Street
University of Illinois
Urbana, IL 61801 USA

ABSTRACT

Nonlinear solutal convection in a melt with a planar solidifying surface and subjected to rotational effects due to both Coriolis and centrifugal forces was investigated in the limit of small segregation. An evolution equation was derived governing the cellular structure of the flow of a binary alloy. The effects due to both Coriolis and centrifugal forces on the flow of the melt and on the onset of cellular structure were determined asymptotically.

INTRODUCTION AND FORMULATION

The results of the recent First International Workshop on Materials Processing in High Gravity (Dubna, Soviet Union, May 1991) indicated the need for a fundamental under-standing of the fluid dynamics of solidifying melts and, in particular, the effects of Coriolis and centrifugal forces on the melt motion. These indications were reinforced by the results presented at the Second International Workshop on Materials Processing in High Gravity (Potsdam, USA, June 1993). The present study includes the effects of both Coriolis and centrifugal forces, and is an extension of the zero–rotation case[1] and the rotating case with zero–centrifugal force.[2,3] We find some interesting results. In particular, we find that, in contrast to the zero–centrifugal force case, the presence of the centrifugal force can stabilize the melt motion.

We consider the problem of solutal convection in a horizontal layer of a binary alloy melt, rotating about a vertical axis with constant rotation rate, Ω, from which a semi-infinite slab of crystal is being grown. The governing system of equations and boundary conditions is considered in a coordinate system rotating about the vertical z–axis with rotation rate Ω and translating in the vertical direction with the velocity V_0 of the solidification form, which is at $z = 0$. This model is the same as that for a planar interface,[2] except for the addition of centrifugal force (per unit mass) term of the form $\beta \Omega^2 C^*(x\hat{i}+y\hat{j})$, where β is the fractional change in density due to a change in solute concentration C^*, (x,y) are the horizontal variables, and $(\hat{i},\hat{j})$ are the unit vectors in the x and y directions, respectively. The details of the full system in the absence of the above force term are given elsewhere.[2,3] We assume that the segregation coefficient k is small ($k \ll 1$), and that crystal growth is proceeding vertically downward (in the positive z direction), so that the solidification generates a thermally destabilizing density gradient. The expression for the unperturbed solute concentration C_0 is given elsewhere[2,3] and will not be repeated here. As is customary,[1-6] we have neglected the thermal contribution to the density gradient, since the solute contribution in the present problem dominates over that of temperature.

It is convenient to use nondimensional variables in which lengths, velocities, time, and solute concentration in the melt motion are scaled by D/V_0, V_0, D/V_0^2, and $(1-k)\, C^\infty/k$, respectively. Here, C^∞ is C^* in the bulk of the melt and D is the diffusion coefficient. Taking the vertical components of curl and double curl of the momentum equation in the limit of infinite Schmidt number, leads us to a relatively simple system of three equations for $C = C^*-C_0$, V and Ψ. Here, V and Ψ are the poloidal and toroidal components of velocity vector u, given by $u = \nabla \times \nabla \times \hat{k}V + \nabla \times \hat{k}\, \Psi$, where $\hat{k}$ is a unit vector in the vertical direction. The simplification of the limit of infinite Schmidt number is assumed here since the results for the regime of practical interest indicate that such a limit is expected, so long as the Schmidt number is greater than about 10.[5]

The above governing system contains three nondimensional parameters, R, τ, and s. Here, $R = g(1-k)\beta C^\infty D^2/(kvV_0^3)$ is the solutal Rayleigh number, g is acceleration due to gravity, v is the kinematic viscosity, $\tau = 2D^2\Omega/(v\, V_0^2)$ is the Coriolis parameter (square root of Taylor number), and $s = \beta(1-k)C^\infty\Omega^2 D^3/(kvV_0^4)$ is a new parameter, which is called the centrifugal parameter and represents the centrifugal force effects. We shall discuss the method of solution to the governing system using a multiscale procedure similar to those used previously.[1-3]

ANALYSIS AND RESULTS

For the weakly nonlinear behavior of the system for small k and s, we let:[1-3]

$$(k,s) = \epsilon^2(k_1,s_1),\ R=R_0(1+\epsilon\mu),\ (x',y') = \epsilon^{1/2}(x,y),\ (z',C') = (z,C)\ , \tag{1}$$
$$(V',\Psi') = \epsilon(V,\Psi),\ t' = \epsilon^2 t\ ,$$

where $0 < \epsilon \ll 1$. We use these rescaled variables and parameters, and then drop the primes. Next, we expand the dependent variables in powders of ϵ:

$$(v,\Psi,C) = \sum_{m=1} \epsilon^m (v_m,\Psi_m,C_m) \tag{2}$$

To order ε, the governing system of equations yields zero solution for V_1 and Ψ_1, and:

$$C_1 = A \exp(-z) \tag{3}$$

where the amplitude of the concentration perturbation, A, is a function of x, y, and t. To order ε^2, the governing system of equations yields solutions for V_2, Ψ_2, and C_2 the same as those given elsewhere,[2,3] and will not be repeated here. The solvability conditions for the ε^2–order system yields:

$$R_0 = 2(1 + \tau^2)\left[1 + (2\tau)^{1/2} + \tau\right]/\left[1 - \tau + \tau(2\tau)^{1/2}\right] \tag{4}$$

As expected, we found that R_0 increases with τ. For $\tau \to 0$, the expression for R_0 approaches the value 2, in agreement with the critical R computed in the limit $k \to 0$, and infinite Schmidt number by Hurle *et al.*[5] for the nonrotating problem. The expression for R_0 tends to the asymptotic limit:

$$R_0 \to \tau(2\tau)^{1/2} \quad \text{as} \quad \pi \to \infty \tag{5}$$

To order ε^3, the solutions for V_3 and Ψ_3 are determined and used in the solvability condition to determine the following evolution equation for A in the two–dimensional case A(x,t):

$$k_1 A + A_t + h_0 A_{xxxx} + h_1 A_{xx} + M(\tau)S\ (2A + xA_x) - h_2(A_x^2 + AA_{xx}) = 0 \quad , \tag{6}$$

where

$$M(\tau) = \left[\frac{(2\tau)^{1/2}}{\tau + 1 + (2\tau)^{1/2}}\right]/(1 + \tau^2) \quad , \tag{7}$$

The expressions for the coefficients $h_n (n=0,1,2)$ (functions of τ and μ) are lengthy and will not be given here, and $A_x = (\partial A)/(\partial x)$, etc. The expression for $M(\tau)$ given by Eq. (7) is non–negative, $M=1/2$ at $\tau=0$, $M \to 0$ as $\tau \to \infty$ and $M(\tau)$ decreases with increasing τ. Next, we reduce the number of coefficients in Eq. (6), first by choosing μ such that $h_1 = h_2$ (h_1 is proportional to μ). This fixes R and defines ε. We found that μ is positive and increases with τ. Then we set:

$$k_1 = k_{\text{eff}_1} h_2^2/h_0 \ , \quad t = T h_0/h_2^2 \ , \quad x = \xi(h_0/h_2)^{1/2} \ , \quad M = N h_2^2/h_0 \quad . \tag{8}$$

Using these rescalings, Eq. (6) yields:

$$\left(2\,Ns_1 + k_{\text{eff}_1}\right)A + A_T + A_{\xi\xi\xi\xi} + A_{\xi\xi} - \left(A_\xi^2 + AA_{\xi\xi}\right) + s_1 N(\tau)\xi A_\xi = 0 \quad . \tag{9}$$

For $\xi \to 0$ (or for $s_1 N \to 0$), Eq. (9) is the same equation as for the nonrotating case,[1] where $k_{eff} = 2Ns_1 + k_{eff_2}$ equals a constant times k_1. Here, we find that k_{eff_2} increases with τ, k_{eff} increases with s_1, and k_{eff} is larger due to the centrifugal force. Also, $N(\tau)$ decreases with increasing τ. Thus, rotation leads to an effective segregation coefficient k_{eff} given by:

$$k_{eff} = (2sMh_0 + kho)/(\epsilon^2 h_2^2) \quad , \tag{10a}$$

$$\epsilon = (R - R_0)/R_0 \mu \tag{10b}$$

larger than the physicochemical value k by an amount determined by the rotation parameters τ and s. Young and Davis[7] considered Eq. (9) in the nonrotating case ($\tau=s=0$), subject to periodic boundary conditions in the interval $0 \le \xi \le L$. They solved Eq. (9) using an explicit finite difference scheme for various values of L chosen as multiples of the wavelength $2\pi\sqrt{2}$, which gives maximum amplification in the linear theory. This linear theory gives $k_{eff} < 1/4$ for instability. Figure 7 in Ref. 7 shows the calculated A for $k_{eff}=0.2$, 0.24, and 0.4. When $k_{eff} > 1/4$, initial disturbances decay to zero and the zero basic state value (A=0) is regained. For $k_{eff} < 1/4$, the system is unstable and a cellular structure forms. For $k_{eff} \ll 1/4$, the results reported in Ref. 7 indicate an apparent secondary instability where the tips of the cells show a tendency to split. Since Eq. (9) has the same form as the corresponding one in the nonrotating case in the limit $\xi N s_1 \to 0$, we see from the above results that in such limit the presence of rotation can stabilize the system and inhibit the onset of a cellular structure by effectively magnifying the segregation coefficient. An important difference between the effects of Coriolis and centrifugal forces is that $|\mu|$ and $|C|$ increase with s, while $|\mu|$ and $|C|$ decrease with increasing τ.

For $s_1 \ne 0$, Eq. (9) admits a steady solution of the asymptotic form:

$$A = (k_{eff} + 2Ns_1)/(6\xi^2) , \quad \xi \gg 1 \quad . \tag{11}$$

Linear stability of infinitesimal perturbation $\bar{A}_1(\xi)\exp(\sigma t)$ about the solution in Eq. (11) leads to the result that:

$$\bar{A} \propto \xi^{-n} \, (0 < n \ll 1) , \quad \xi \gg 1 \quad , \tag{12a}$$

$$\sigma = -2/3 \left(k_{eff_1} + Ns_1\right) < 0 \quad . \tag{12b}$$

Hence, Eq. (11) is stable. For $s_1=0$, Eq. (9) admits weakly nonlinear periodic solutions that turn out to be unstable with respect to infinitesimal disturbances.

CONCLUSIONS

1. The critical Rayleigh number R_0 increases with Coriolis parameter. It is in agreement with zero-rotation results[1,6] in the limit of $\tau \to 0$.

2. Rotational effects lead to an effective segregation coefficient larger than the equilibrium value for $\xi N s_1 \to 0$.

3. For very strong Coriolis force ($\tau \to \infty$), centrifugal effect is insignificant ($sM \to 0$).

4. For negligible Coriolis force ($\tau \to 0$), the centrifugal force has a stabilizing effect for $\xi \to 0$.

5. The effect of the centrifugal force, due to term ξA_ξ in Eq. (9), is to make the solution for solute and velocity nonperiodic and noncellular, even at the linear level.

6. For zero centrifugal effect, the weakly nonlinear solution is cellular and periodic solutions are permitted.

7. The gravity effect is destabilizing for $\xi N s_1 \to 0$ in the sense that the effective segregation coefficient decreases with increasing g.

8. The flow velocity and the solute concentration decrease with increasing τ and decreasing s.

9. For sufficiently large $\tau(\tau \to \infty)$, convection is suppressed entirely by rotation. For sufficiently large $s(s \to \infty)$, convection persists.

REFERENCES

1. D.S. Riley and S.H. Davis, Hydrodynamic stability of the melt during the solidification of a binary alloy with small segregation coefficient, *Physica D* 39:231 (1989).
2. D.N. Riahi, Directional solidification in a rotating system with small segregation coefficient, *Phys. Rev. B* 44:4170 (1991).
3. D.N. Riahi, Effect of rotation on the stability of the melt during the solidification of a binary alloy, *Acta Mech.*, in press.
4. D.T.J. Hurle, E. Jakeman, and A.A. Wheeler, Effect of solutal convection on the morphological stability of a binary alloy, *J. Crystal Growth* 58:163 (1982).
5. D.T.J. Hurle, E. Jakeman, and A.A. Wheeler, Hydrodynamic stability of the melt during solidification of a binary alloy, *Phys. Fluids* 26:624 (1983).
6. D.N. Riahi, Solutal convection in the melt during solidification of a binary alloy, *Phys. Fluids* 31:27 (1988).
7. G.W. Young and S.H. Davis, Directional solidification with buoyancy in systems with small segregation coefficient, *Phys. Rev. B* 34:3388 (1986).

MODELING AND EXPERIMENTS ON EPITAXIAL
GROWTH ON A GaAs HEMISPHERE SUBSTRATE
AT 1 g AND UNDER HYPERGRAVITY

Jean-Claude Launay,[1,2] Stéphanie Bouchet,[1]
Anthony Randriamampianina,[3] Patrick Bontoux[3] and
Pierre Gibart[4]

[1]PRAME - Aërospatiale
 B.P. 11
 F-33125 St.Médard-en-Jalles Cédex, France
[2]Laboratoire de Chimie du Solide du CNRS
 Université de Bordeaux 1
 351 Avenue de la Libération
 F-33405 Talence Cédex, France
[3]Institut de Mécanique des Fluides - CNRS
 1 Rue Honorat
 F-13003 Marseille, France
[4]Laboratoire de Physique du Solide et Energie et Energie Solaire - CNRS
 Parc Sophia Antipolis
 F-06560 Valbonne, France

ABSTRACT

Centrifuge chemical vapor transport experiments under 5 and 10 g produced growth of GaAs on [001] GaAs substrate oriented with a macrostructure typical of diffusion-controlled growth. A current hypothesis for this phenomenon is that Coriolis and gravity gradient forces produced by the centrifugal motion can effectively damp buoyancy-driven convective flows.

Numerical simulation using spectral methods was carried out for axisymmetric flow regimes, and was compared to experiments.

INTRODUCTION

For almost three decades, GaAs single crystals and epitaxial layers have been grown from the vapor phase using HCl as a transport agent. This is a well documented system, and the thermodynamics and kinetics are very well understood.

Expitaxial growth from the vapor (in an open system) is still used for the fabrication of GaAs–based devices like light emitting diodes.

When analyzing the growth velocity as a function of temperature, two main temperature ranges should be distinguished:

(i) At high temperatures (T ≥ 800 °C) the growth rate is controlled by mass transport only and can be deduced in a straightforward way from basic thermodynamics and flow parameters.

(ii) A low temperature region where growth is kinetically controlled, e.g., the growth rate depends on the nature of the crystal faces. This has been discussed in detail by Shaw,[1] Cadoret[2] and Hollan.[3] Cadoret,[4] in a deep analysis of the growth mechanisms, showed that GaAs incorporation in the growing layer is controlled by a GaAs–Cl surface complex.

In the present study, epitaxial GaAs was grown in closed system on GaAs single crystal half–spheres in the kinetically controlled temperature range, under different g conditions:

(i) Microgravity (EURECA I experiments).

(ii) 1 g and hypergravity. The present paper reports on 1 g and hypergravity experiments and models the mass transport under these conditions.

Since, the buoyancy–driven regime is recognized to play an important role for the understanding of mass transport and crystal growth phenomena, much work has been done to find the appropriate experimental conditions that would dampen the perturbations induced by convection. The very low gravity level offered by space is ideal, although not free from many important difficulties. Experiments on vapor transport of GaAs were carried out on board the platform EURECA–1. We selected the GaAs:HCl system for the investigation of the effects of high acceleration on vapor transport and crystal growth properties, because it is a well documented system. Chemical vapor transport experiments of GaAs in the presence of a gaseous mixture 3% HCl, H_2 were performed under 1, 5 and 10 g acceleration, with stabilizing and destabilizing conditions (g parallel and antiparallel to the gradient ∇T, respectively).

The objective was to determine the influence of gravity on the quality of the crystals, the growth anisotropy and the hydrodynamics.

EXPERIMENTAL SETUP

The GaAs original material used as source and substrate came from a single crystal obtained by the Czochralski method. It was semi–insulating and had a resistivity of 3.7 × 10^7 Ω cm.

The polished substrates were GaAs hemispheres, [001] oriented.

Six g of GaAs were loaded into pretreated fused silica ampoules of 20.8 mm outer diameter, 17.6 mm inner diameter, and 95 or 115 mm length. The loaded ampoules were evacuated at room temperature under a pressure of about 10^{-1} Pa for 24 hr. Afterwards, a gas mixture 3% HCl/H_2 was introduced into the ampoule till a pressure of 0.1 atm at room temperature was obtained. The ampoule was then sealed.

The configuration of the experiment corresponded to two ampoules, placed in a cartridge, in the furnace (Fig. 1). One ampoule was in a temperature range 785 °C to 750 °C (it will be called DT1), whereas the other was at 750 °C to 700 °C (DT3). The temperature profile is given by Fig. 2. These furnaces were then integrated on the centrifuge facility.

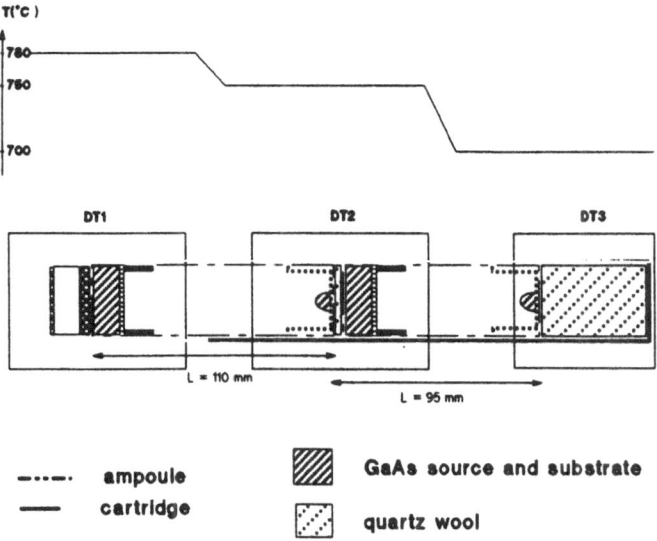

Figure 1. Configuration of the equipment.

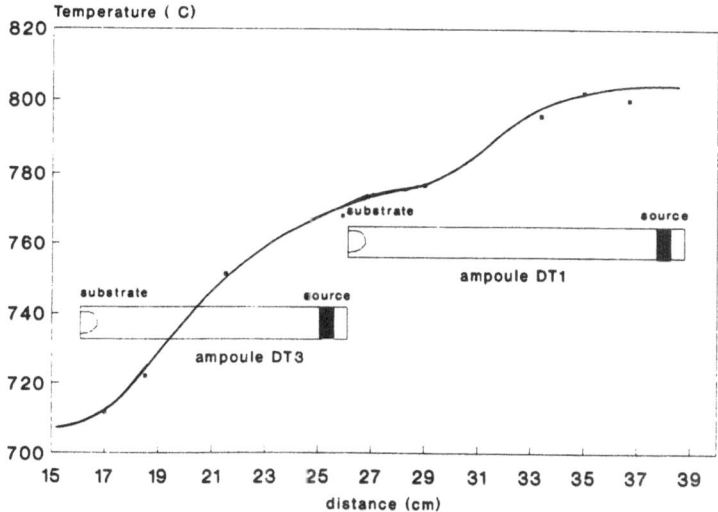

Figure 2. Schematic diagram of the temperature profile and ampoules' locations employed in this work.

Description of the Centrifuge

We used the centrifuge installed at the Nantes Laboratoire Central des Ponts et Chaussées. The radius of the platform is 5.5 m. The maximum centrifugal acceleration is 200 g.

The growth ampoule was integrated either in an horizontal furnace (SAER) or in a vertical furnace (CNES). After the desired temperature profile was established, the required acceleration level was established within a few minutes. The transport and crystal growth periods for the experiments were performed under 5 or 10 g for about 80 hr.

Disposition of the Furnaces on the Basket

5 g. For this gravity level, the furnace was in a horizontal position. The acceleration was calculated at point C on the axis of the furnace (Fig. 3). The distance between point C and the axis of the centrifuge is the reference for the centrifuge gravity vector.

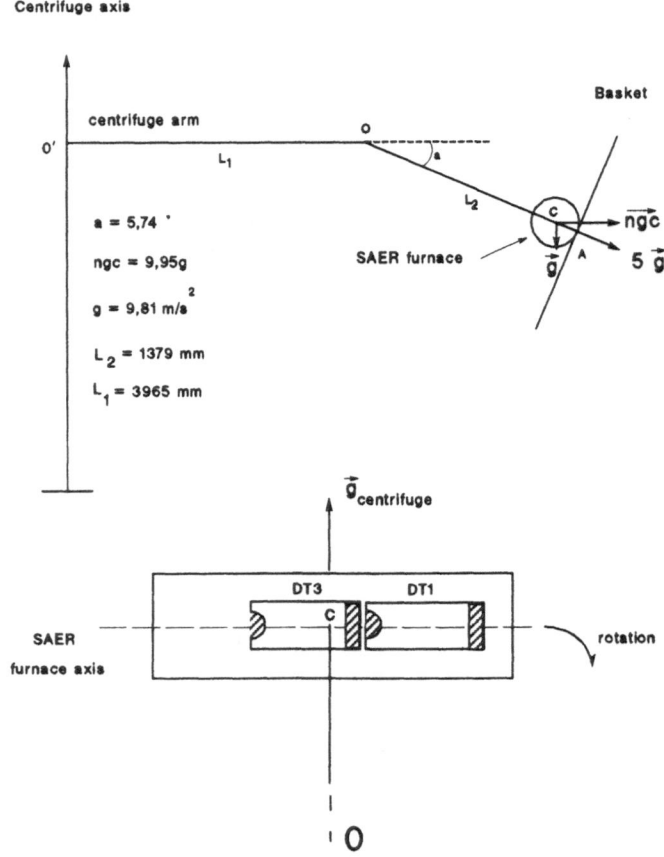

Figure 3. Sketch of the experimental configuration of the ampoules at 5 g.

10 g. In this case, we installed two furnaces on the basket: one was vertical (CNES) and the other was horizontal (SAER). The furnace CNES was in a destabilizing position, that is to say the acceleration vector g was in the opposite direction with respect to the mass flux. Figure 4 shows the representation of the two furnaces on the platform. The point C was where the acceleration 10 g was calculated.

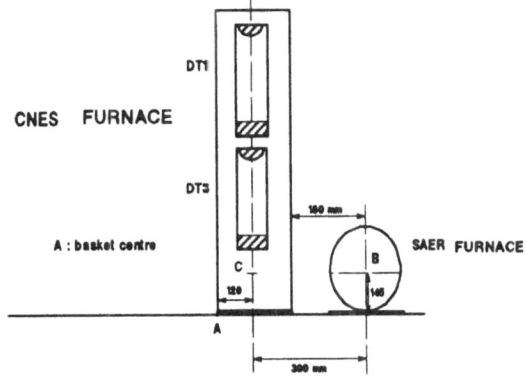

Sketch of the furnaces on the basket for 10 $\vec{g}$

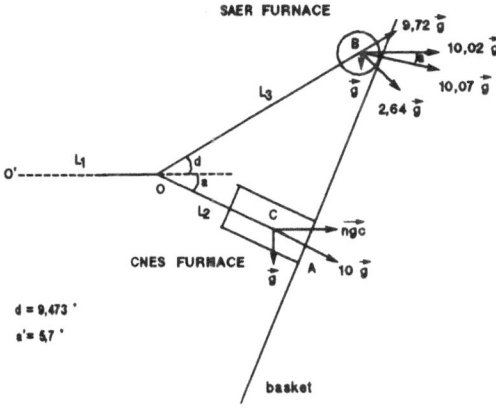

Figure 4. Sketch of g vectors in the 10 g experiment.

It is possible to calculate the resultant accelerations in different directions at different points, at the source and at the substrate, for example. These vectors are reported in Table 1. We mention the variation of the total density of the gaseous phase. We can also deduce the product ρg. Its variation is low, because the gravity gradient ∇·g between the source and the substrate is important in each growth ampoule.

THERMODYNAMIC BACKGROUND

In the temperature range considered, 800-1200 K, the major species are GaAs(s), GaCl(g), GaCl$_3$(g), HCl(g), H$_2$(g), Cl$_2$(g) and As$_4$(g). (GaCl$_2$(g) and AsCl$_3$(g) can be considered as minor products.) The following reactions describe the equilibria between the species involved in the transport of gallium arsenide by chlorine in the presence of hydrogen.

$$2\,GaA_s + 2\,HCl_g = 2\,GaCl_g + H_{2g} + 0.5\,As_{4g} \tag{1}$$

Table 1. Resulting accelerations in DT1 and DT3 ampoules.

SAER Furnace

	Temperature (K)	Centrifuge vector (m.s^{-2})	Resulting gravity vector (m.s^{-2})	Density (g/m^3)	Product ρg (g.m^2.s^{-2})
source	1023.15	10.023	10.073	29.57	298
DT3					
substrate	973.15	10.03	10.079	31.09	313
source	1058.15	10.036	10.086	29.99	302
DT1					
substrate	1023.15	10.035	10.085	31.02	313

CNES Furnace

	Temperature (K)	Centrifuge vector (m.s^{-2})	Resulting gravity vector (m.s^{-2})	Density (g/m^3)	Product ρg (g.m^{-2}.s^{-2})
source	1023.15	9.58	9.64	29.57	285
DT3					
substrate	973.15	9.41	9.46	31.09	294
source	1058.15	9.79	9.84	29.99	295
DT1					
substrate	1023.15	9.62	9.67	31.02	300

$$GaCl_{3g} = GaCl_g + Cl_{2g} \tag{2}$$

$$2\,HCl_g = H_{2g} + Cl_{2g} \tag{3}$$

It had been shown[4] that the system GaAs:H:Cl can be fully described using only these three equilibria.

Gas Phase Composition

The partial pressures of the gas species at equilibrium were estimated by the software GEMINI 1 (Gibbs Energy Minimizer). It allowed us to calculate the partial pressure of each gaseous species in equilibrium with solid GaAs by minimization of the total Gibbs energy of the system under either constant pressure or volume conditions. The free energies of formation of GaAs(s), GaCl(g), GaCl$_3$(g), HCl(g), H$_2$(g), Cl$_2$(g) and As$_4$(g) were taken from the data bank THERMODATA.[5] The temperature dependence of the partial pressures is represented in Fig. 5. P_{H_2} and P_{tot} are not mentioned because they are higher than the others (about 0.3 atm). The gas phase composition for the DT1 and DT3 ampoules is given in Table 2.

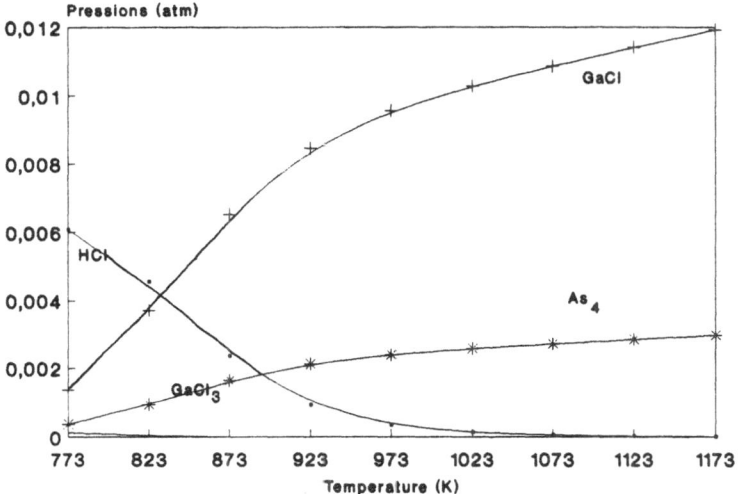

Figure 5. Partial pressures of gaseous species versus T.

Supersaturation

To evaluate the supersaturation we suppose that the total pressure is constant throughout the tube, the variation of the partial pressures between the source and the substrate is linear, and the regime is diffusional ($P_{tot} < 4 \times 10^{-1}$ atm). The supersaturation is the difference between the local partial pressure $P(y)$ and the equilibrium vapor pressure $P(Ty)$ corresponding to the local temperature $T(y)$. We define γ via:

$$1 + \gamma(y) = P(y)/P(Ty) \tag{4}$$

with

$$P(y) = P(Tp) = \left(P(Ts) - P(Tp)\right) y/L \tag{5}$$

where L is distance between the source and the sink, Ts is source temperature, Tp is sink temperature, and y is position in the ampoule.

Table 2. Gas phase composition - $P(GaCl_3)$, $P(Cl_2)$, $P(AsCl_3) < 10^{-8}$ atm.

Ampoule DT1

Temperature (K)	Vapor pressure (atm)	GaCl pressure (atm)	As$_4$ pressure (atm)	HCl pressure (atm)	H$_2$ pressure (atm)
1023.15	0.365	1.027×10^{-2}	2.568×10^{-3}	1.307×10^{-4}	0.3520
1038.15	0.370	1.045×10^{-2}	2.612×10^{-3}	1.072×10^{-4}	0.3573
1058.15	0.376	1.068×10^{-2}	2.671×10^{-3}	7.591×10^{-5}	0.3775

Ampoule DT3

Temperature (K)	Vapor pressure (atm)	GaCl pressure (atm)	As$_4$ pressure (atm)	HCl pressure (atm)	H$_2$ pressure (atm)
973.15	0.347	9.546×10^{-3}	2.386×10^{-3}	3.437×10^{-4}	0.3345
998.15	0.356	9.906×10^{-3}	2.476×10^{-3}	2.372×10^{-4}	0.3432
1023.15	0.365	1.027×10^{-2}	2.567×10^{-3}	1.306×10^{-4}	0.3518

In the present case:

$$1 + \gamma = \frac{P_{As_4}^{0.25} \, P_{GaCl} \, P_{H_2}^{0.5}}{P_{HCl}} \cdot \frac{P_{HCl_{eq}}}{P_{As_{4\,eq}}^{0.25} \, P_{GaCl_{eq}} \, P_{H_2\,eq}^{0.5}} \tag{6}$$

The supersaturation profiles calculated for the temperature profile are given in Fig. 6. The maxima of the curves locates the nucleation and the growth of GaAs; here, at about 2.0 and 3.0 mm from the cold extremities of the ampoules.

Mass Flow

In a closed tube, the gaseous species are transported from the hot source to the substrate by diffusion due to the concentration gradient, and laminar flow due to the molar volume changes accompanying the chemical reactions. Convection and kinetic limitations are neglected. The total flux of the ith gas species is given by the sum of Stephan's flux and Fick's flux:

$$J_i = UP_i/RT - (D_i/RT) \cdot dP_i/dy \tag{7}$$

where U is flow velocity (cm.s^{-1}), D is diffusion coefficient (cm^2.s^{-1}), P is partial pressure (atm) and U is the mean drift velocity imposed on the molecules by the pressure difference.

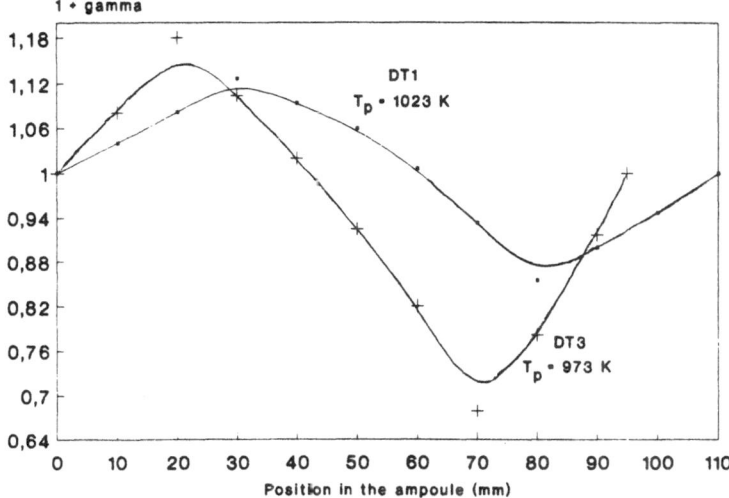

Figure 6. Supersaturation profile along DT1 and DT3 ampoules.

The binary diffusion coefficients are estimated from the formula given by Reid and Sherwood:[6]

$$D_{ij} = (M_i + M_j)^{0.5} \cdot (1.858 \times 10^{-3} \cdot T^{1.5})/\left(P_{tot} \cdot \sigma i_j^2 \cdot \Omega_D\right) \cdot (M_i \cdot M_j)^{0.5} \qquad (8)$$

where P_{tot} is total pressure (atm), σ_{ij} is force constant of the binary ij, Ω_D is the collision integral, M is molecular weight (g/mol) and T is temperature (K).

The diffusion coefficient of component i for a multicomponent vapor is estimated by:

$$D_i = (1 - y_i) \text{ SCALESYM100}/^{j\neq i} \text{ SCALESYM300}\sum_{j=1}^{n} \left(\frac{y_j}{D_{ij}}\right) \qquad (9)$$

where y_i is the mole fraction of i in the gas.

It is also possible to calculate the total flux. The theoretical values are on the order of 19×10^{-9} mol cm^{-2} s^{-1} for ampoule DT1 and 35×10^{-9} mol cm^{-2} s^{-1} for ampoule DT3. We have carried out this calculation assuming that there are no adsorption or nucleation barriers at the source and seed interfaces.

Table 3 shows the values of experimental mass flux obtained after growth. On average, in the horizontal configuration, that is to say g perpendicular to the ampoule axis, the flux of the 1 g experiments in ampoule DT1 was about 3.5×10^{-9} mol cm^{-2} s^{-1} and in DT3 was 6×10^{-9} mol cm^{-2} s^{-1}. The ratio between theoretical and experimental mass flux is about 5.

It should be pointed out that for 5 g and 10 g, increases in the acceleration by a factor of 5 and 10 increased the mass flux by a factor of about 1.9 and 1.95, respectively (Table 3). Thus, the mass flux varied with acceleration as $g^{0.5}$ for 5 g and $g^{0.25}$ for 10 g.

Table 3. Experimental flux.

| Gravity level | Flux (10^{-9} mol cm^{-2} s^{-1}) | | |
	DT1	DT3	
1 g	3.5	6.0	no doping
		0.9	Sn doping
5 g	6.7		no doping
		0.4	Sn doping
10 g	6.8	8.8	no doping

Faceting of the Half Sphere Substrate

After removal of the ampoule from the cartridge, photographic documentation of the ampoule and deposition patterns was obtained. Photographs of the closed ampoules show a symmetrical crystal deposition pattern about the axis of the furnace (Fig. 7). However, when g was increased from 5 g to 10 g, the parasitic crystal deposition on the ampoule inner wall became more important.

After the growth ampoules were opened, we noted that facets had appeared on the half spheres. These facets were symmetrically arranged to the axis of the furnace (Fig. 8).

Figure 8a shows the morphology of the [100] GaAs half sphere after an 80 hr epitaxial growth run. Several facets appeared on the surface, and were identified by Läue back-diffraction photographs. A (100) facet appeared on the top. In addition, (113), (11$\bar{3}$), (110), and (111) Ga facets were identified.

Figure 8b shows a nonoriented half sphere after an 80 hr growth run under 10 g. The three triangular facets are {110}, and the irregular facet in the center is a {111} Ga face.

A much more detailed analysis is achievable after cutting the half spheres along one well defined diametral plane, polishing this surface, and etching with the classical A–B mixture. In this way, a perpendicular intersection with several well defined planes is obtained. Under SEM examination, the A–B etch revealed the interface between the initial GaAs seed and the GaAs epitaxial layer, and allowed a direct evaluation of the growth velocities of the identified planes.

Figure 9 represents the deposition rate in μm/h versus the angle with the (001) plane. Planes with gallium polarity exhibit the highest growth velocity. We notice an asymmetry about the (001) plane. These results are in perfect agreement with previous papers.[1,4,7]

The growth velocities are higher at 973 K than at 1028 K, except in the case of the (001) plane. This feature could be explained by the curve $\tau = f(l/t)$ obtained by Gentner.[7] He observed that the growth velocity increases from a critical temperature when the substrate temperature is decreased. This critical temperature depends on the value of the partial pressure of the GaCl active species. These results explain the higher growth rate of the DT3 ampoule (seed at 973 K) in comparison with the DT1 ampoule (seed at 1023 K) at $P_{GaCl} = 10^{-3}$ atm.

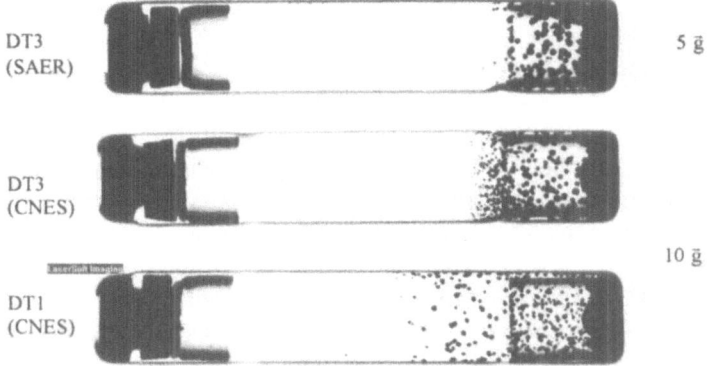

Figure 7. X–ray photographs.

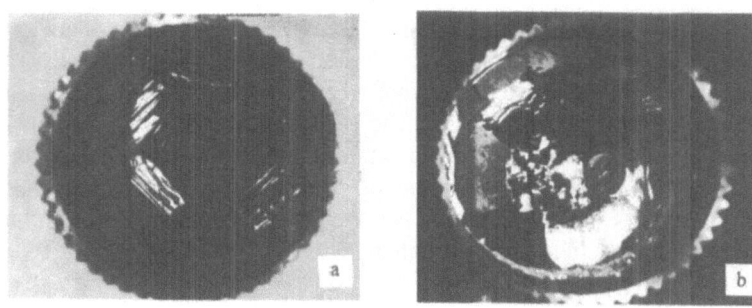

Figure 8. Morphology of the GaAs half sphere after epitaxial growth run.

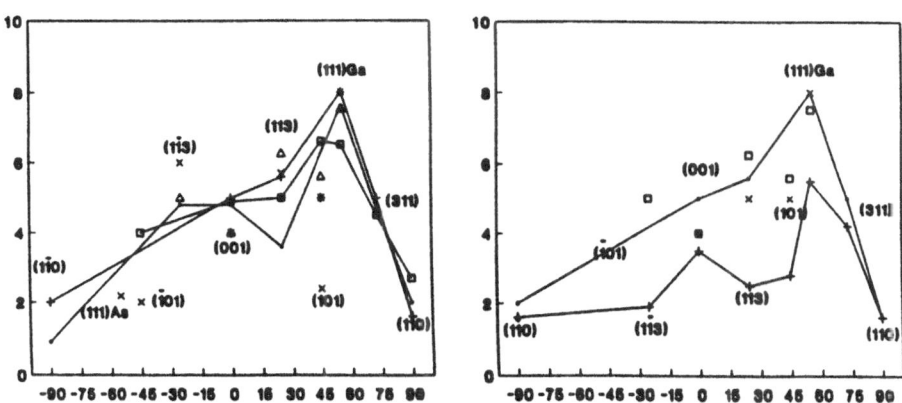

Figure 9. Growth rate (μm/h) deviation from (001) orientation.

NUMERICAL MODELING

Rotation can generate various kinds of complex fluid behavior. In particular, Stewartson and Ekman layers can develop in the vicinity of the walls. The computation of such flows requires high resolution, especially near the walls, if the purpose is to adequately study the transition to time-dependent motions as observed in experiments. Spectral methods based on Chebyshev expansion are well adapted to the prediction of boundary layer motion, since the Chebyshev (Gauss-Lobatto) collocation points are concentrated near the boundaries of the computation domain. Moreover, the motivation for the use of spectral methods stems from the attractive approximation properties of orthogonal polynomial expansions, particularly the exponential convergence, which is referred to as spectral accuracy. These techniques have become very popular since the pioneering works of Gottlieb and Orszag,[8] who made the first unifying mathematical assessment of the theory.

More recently, a complete mathematical study of spectral techniques was developed by Canuto *et al.*[9] For spectra-tau methods, the solution is calculated in spectral space, and a suitable orthogonality condition ensures that the residue is minimized. This requires Fast Fourier transforms for the transfer between the physical and spectral spaces.

The tau-method was used to study the competition between buoyancy and rotation forces in a differentially-heated rotating annulus for large curvature parameter and aspect ratio, similar to the experiments. The ampoule was modeled by a rectangular section of an annular domain of height 2H extending from -H to H in the z-direction and from R_I to R_E in the r-direction, with $R_E - R_I \ll R_I$. The two vertical cylinders are held at constant temperatures, T_I and T_E. Different temperatures profiles are imposed on the horizontal walls, according to the experiments considered. The axis of rotation is oriented antiparallel to the gravity vector. Convective motion is driven by the buoyancy force and modified by the variable acceleration field and by rotation.

Governing Equations

The governing equations are written for a Newtonian fluid with the Boussinesq approximation. The density is taken to be constant except in the effective body force terms (buoyancy, Coriolis, and centrifugal), as suggested by Homsy and Hudson,[10] De Vahl Davis,[11] and Randriamampianina.[12] When a linear variation with temperature is assumed:

$$\rho = \rho_0 \{1 - \beta (T - T_0)\} \tag{10}$$

where $T_0 = (T_E + T_I)/2$, ρ_0 and T_0 are the reference density and temperature, and β is the thermal expansion coefficient.

The Navier-Stokes and energy equations are expressed in a vorticity-stream function formulation for axisymmetric flow regimes. This formulation eliminates the cumbersome treatment of the pressure, and also ensures the continuity equation to be exactly satisfied irrespective of the resolution.

The space variables r' and z' are made dimensionless using $\Delta R/2$ and H, in a way to have the computing domain inside the square $[-1,1] \times [-1,1]$: $r' = (2/\Delta R)r - R_m$ and $z' = z/H$, where $\Delta R = R_E - R_I$ and $R_m = (R_E + R_I)/\Delta R$ is the curvature parameter. The temperature is normalized as $\theta = 2(T - T_0)/\Delta T$, with $\Delta T = T_E - T_I$. The velocity scale is taken as K/H, where K is the thermal diffusivity.

The dimensionless equations are expressed in a rotating frame as:

$$\Delta\Psi = \left(\frac{r + R_m}{R_m}\right)\zeta \tag{11}$$

$$\frac{\partial\zeta}{\partial t} + V\cdot\nabla\zeta = \frac{R_a P_r}{16\,AR^4}\frac{\partial\theta}{\partial r} -$$

$$\left\{\frac{2\,R_a\,F_r}{R_e^2\,P_r}\frac{v^2}{AR(r+R_m)} + \frac{R_a\,F_r\,v}{2\,R_e\,AR^2} + \frac{R_a\,F_r\,P_r}{32\,AR^3}(r+R_m)\right\}\frac{\partial\theta}{\partial z} \tag{12}$$

$$+ \left\{\frac{2\,v}{AR(r+R_m)} + \frac{R_e\,P_r}{4\,AR^2}\right\}\frac{\partial v}{\partial z} + P_r\,\nabla^2\zeta$$

$$\frac{\partial v}{\partial t} + V\cdot\nabla v = P_r\,\nabla^2 v + \frac{2\,R_a\,F_r}{R_e^2\,P_r\,AR}\frac{u\,v\,\theta}{R_m+r} + \frac{R_a\,F_r}{2\,R_e\,AR^2}u\,\theta - \frac{R_e\,P_r}{4\,AR^2}u \tag{13}$$

$$\frac{\partial\theta}{\partial t} + V\cdot\nabla\theta = \nabla^2\theta \tag{14}$$

where:

$$\Delta = \frac{1}{AR^2}\frac{\partial^2}{\partial r^2} + \frac{\partial^2}{\partial z^2} - \frac{1}{AR^2(r+R_m)}\frac{\partial}{\partial r}$$

$$V\cdot\nabla = \frac{u}{AR}\frac{\partial}{\partial r} + w\frac{\partial}{\partial z} + \frac{u\delta_1}{AR(r+R_m)}$$

$$\nabla^2 = \frac{1}{AR^2}\frac{\partial^2}{\partial r^2} + \frac{\partial^2}{\partial z^2} + \frac{1}{AR^2(r+R_m)}\frac{\partial}{\partial r} - \frac{\delta_2}{AR^2(r+R_m)^2}$$

$\delta_1 = -1$ for ζ; $\delta_1 = 1$ for v; $\delta_1 = \delta_2 = 0$ for θ; and $\delta_2 = 1$ for ζ and v. The geometry is defined by the aspect ratio $AR = \Delta R/2H$ and the curvature parameter R_m. The physical parameters are characterized by the following dimensionless groups:
- the Prandtl number, $P_r = \nu/K$, where ν is the kinematic viscosity;
- the rotational Reynolds number, $R_e = 2\Omega\Delta R^2/\nu$;
- the Froude number, $F_r = \Omega^2\Delta R/g$; and
- the Rayleigh number, $R_a = g\beta\Delta T\,\Delta R^3/\nu$.

The velocity components in the meridional plane and the vorticity are given by:

$$u = \frac{R_m}{r + R_m} \frac{\partial \Psi}{\partial z} \qquad (15)$$

$$w = -\frac{1}{AR} \frac{R_m}{(r + R_m)} \frac{\partial \Psi}{\partial r} \qquad (16)$$

$$\zeta = \frac{\partial u}{\partial z} - \frac{1}{AR} \frac{\partial w}{\partial r} \qquad (17)$$

The dynamical boundary conditions are the no-slip and no-permeability conditions, $\Psi = \partial\Psi/\partial n = 0$, and $v = 0$ for rotating walls. The temperature profile on the horizontal walls will be defined later depending on the experiment considered.

Numerical Approach

The numerical approach is based on the extension of a spectral code developed by Randriamampianina[13] for the study of flows inside a differentially-heated rotating annular domain. The solution is obtained by using Chebyshev polynomial expansions in two space variables. Thus, the flow variables $\phi = \{\zeta, v, \Psi, \theta\}$ are approximated as:

$$\phi(r,z,t) \approx \phi_{NM}(r,z,t) = \sum_{n=0}^{N} \sum_{m=0}^{M} \phi_{nm}(t)\, T_n(r)\, T_m(z) \qquad (18)$$

where $T_n(r)$ and $T_m(z)$ are the Chebyshev polynomials defined in the square $[-1,1] \times [-1,1]$ by $T_n(r) = \cos[n \cos^{-1} r]$ and $T_m(z) = \cos[m \cos^{-1} z]$, respectively. The coefficients in spectral space $\phi_{nm}(t)$ are the unknowns that contain the time-dependence of the solution.

The residue due to the truncated expansion is set to zero by imposing an orthogonality condition with respect to the basis of the Chebyshev polynomials. This results in the following system of $(N-1) \times (M-1)$ differential equations for each component ϕ_{nm}, $0 \le n \le N-2$ and $0 \le m \le M-2$:

$$\frac{d\phi_{nm}}{dt} = -C\phi_{nm} + A\phi_d D\phi_{nm} + S\phi_{nm} \qquad (19)$$

for the transport equations and:

$$D X \Psi_{nm} = \left[\frac{r + R_m}{R_m}\right] \zeta_{nm} \qquad (20)$$

for Eq. (11) with $DX\Psi = \Delta\Psi$. Here, $C\phi = V \cdot \nabla\phi$, $D\phi = \nabla^2\phi$, and $S\phi$, respectively, correspond to the Chebyshev polynomial expansion of the known coupling terms and source in the equations for ζ and v. The complementary set of equations, for $(N - 1 \le n \le N) \times (0 \le m \le M - 2)$ and $(0 \le n \le N) \times (M - 1 \le m \le M)$, is obtained from the boundary conditions.

The solution is advanced in time using linear multistep time schemes (Gear,[14] Byrne and Hindmarsh[15]). The method is based on the ODE solver LSODA, which involves a selection of schemes (Adams-Moulton and Backward Differentiation Formula, respectively, for nonstiff and stiff systems), and an automatic adaptation of the time step and switching between schemes according to the stiffness of the problem (Petzold,[16] Hindmarsh[17]).

The matrix diagonalization technique of Haidvogel and Zang[18] extended to an axisymmetric geometry (Randriamampianina[12]), is applied in the r-direction for the solution of the Poisson-type equation [Eq. (20)] on Ψ. The solution is obtained by solving $(M - 1)$ one-dimensional Poisson equations in the z-direction for each eigenvalue. This problem involves the resolution of quasitridiagonal matrices assuming diagonal dominance. The transfer between the spectral and physical spaces is facilitated via the FFT algorithm developed by Temperton[19] for a Cray computer.

Numerical Results

All the computations were done on the CRAY-YMP 2E of the Centre de Calcul Régional de l'IMT at Château-Gombert, Marseille. The coefficients required to run the numerical simulation were evaluated at the mean temperature (998 K for DT3 and 1038 K for DT1) for the gaseous species GaCl considered predominant. The resulting values are reported in the following tables:

	μ (μPo)	k (g cal/cm s K)	ρ (g/cm^3)	Cp (cal/g K)	ΔR (nm)	ΔT (K)
DT3	749.75	307×10^{-6}	30.35×10^{-6}	0.3045	95	50
DT1	781.75	329.5×10^{-6}	30.5×10^{-6}	0.3125	115	35

Thus, we have computed the corresponding values of the dimensionless parameters:

	AR	R_m	P_r	R_a	β
DT3	5.4	115.8	0.744	513.26	1/998
DT1	6.5	95.6	0.745	565.65	1/1038

Some values of Froude and Reynolds numbers expressed as a function of the angular rotation velocity, Ω, are given below:

		1 g $\Omega = 0$	5 g $\Omega = 3$ rad s^{-1}	10 g $\Omega = 4.27$ rad s^{-1}
DT3	F_r	0	0.0864	0.173
	R_e	0	21.94	31.23
DT1	F_r	0	0.1045	0.21
	R_e	0	31	44.07

To allow direct comparison with experiments, the axial and radial dimensions were brought back to $0 \leq r \leq 1$ and $0 \leq z \leq 1$ for the whole cavity.

DT1 and DT3 Cavities

For our DT1 and DT3 ampoules, the computations were carried out using 33×33 Chebyshev polynomials. A convergence to steady flow was assumed as soon as the residue between two successive values of the vorticity was less than 10^{-5}.

The thermal boundary conditions on the horizontal walls were defined similarly to the experiments as shown below.

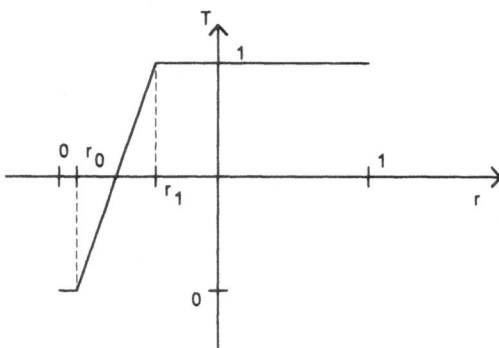

The values of r_0 and r_1 are different for DT1 and DT3. For the two cases, the influence on the flow structure of the rotation speed over fixed thermal gradients was analyzed.

Extremet et al.[20] carried out similar studies for a two-dimensional cavity without rotation and pointed out the effect of the width of the gradient to the aspect ratio. The large value of the curvature parameter R_m, considered for the two present experiments, can bring the cavity back to a two-dimensional configuration. On the other hand, this also enhances the effect of the centrifugal term, as can be seen by inspection of Eq. (12).

DT3 Cavity. The thermal profile on the horizontal walls was defined by a linear variation of temperature between the points $r_0 = 0.010$ and $r_1 = 0.854$ on walls.

When the centrifugal acceleration was increased up to 20 g, no significant change was observed in the isotherms. The configuration obtained for a pure diffuse regime (0 g) is displayed in Fig. 10. Because the Archimède number, $R_a/P_r R_e^2$, which is a measure of the effect of the thermal gradient to rotation, remains large (0.37 for 20 g), the flow is in the buoyancy–driven regime.

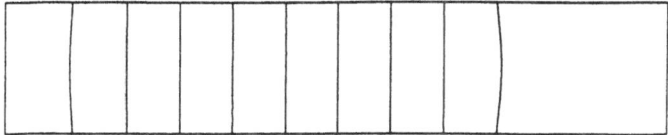

Figure 10. Isotherms for DT3 cavity at 0 g.

We notice that the region where the thermal gradient is dominant occupies almost the whole cavity (85%). This can also be explained by the large curvature parameter bringing the flow back to two-dimensional behavior. This is confirmed by the spread of the thermal cell inside the cavity, as shown by the streamlines displayed for different values of the centrifugal acceleration in Figs. 11a–d. When the cavity is at rest (Fig. 11a), there is symmetry about the mid-plane z = 0.5, with a higher concentration of streamlines towards the inner cylinder due to the asymmetrical location of the thermal gradient.

When increasing the centrifugal acceleration, the streamlines deviate from the upper horizontal wall towards the outer cylinder. This breaking of the symmetry corresponds to the onset of a centrifugal cell, which is counter-rotating to the thermal cell. Then for a value larger than 10 g, another centrifugal cell develops towards the corner of the inner cylinder and the lower horizontal wall.

The axial variations of the radial component of the velocity are displayed in Figs. 12–14 for three values of the centrifugal acceleration, 2.5, 3, and 20 g, respectively. For the lowest value, 2.5 g displayed in Fig. 12, we observed an S-shaped radial velocity at two radial locations, r = 0.5 and 0.8. These are characteristic of a buoyancy-driven flow ascending towards the hot wall (outer cylinder) and descending towards the cold one.

To show more precisely the onset of the centrifugal effect, we have displayed in Fig. 13 the radial velocity profile for 3 g at the first radial node from the outer cylinder, r = 0.98. This velocity profile is for the mid-plane, z = 0.5. The radial velocity profiles at the same radial locations as in Fig. 12 for 2.5 g, but at 20 g, are illustrated in Fig. 14. Two flow regimes coexist for this acceleration. The S-shaped profile, characteristic of a buoyancy-driven regime, is observed at location r = 0.5, while the generation of the second cell due to centrifugal acceleration is clearly visible at r = 0.8, with a flow reversal towards the upper wall.

The radial variations of the axial velocity are reported in Fig. 15 at axial location z = 0.5 for an acceleration of 20 g. We notice an asymmetry with a plateau between the inner and outer cylinders, related to the thermal gradient. The same behavior was observed for all values of the centrifugal acceleration; only the levels change.

DT1 Cavity. In this case, the thermal profile on the horizontal walls is defined by a linear variation of temperature between the points $r_0 = 0.113$ and $r_1 = 0.5$ on the walls. The same comments made for cavity DT3 can be globally applied for this configuration. However, because the thermal gradient is localized to the first half of the cavity, the second half behaves as a "solid." The isotherms are unchanged for all values of the centrifugal acceleration considered (up to 20 g), and are displayed in Fig. 16 for 0 g.

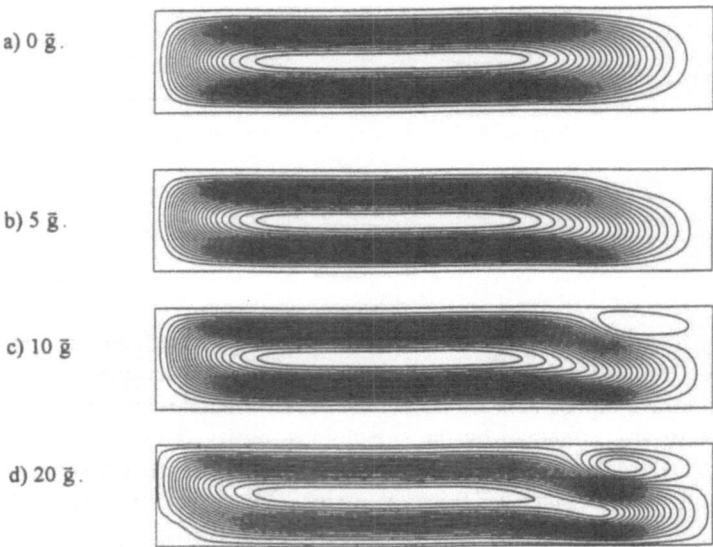

a) 0 ḡ.

b) 5 ḡ.

c) 10 ḡ

d) 20 ḡ.

Figure 11. Flow configurations at different levels of the centrifugal acceleration.

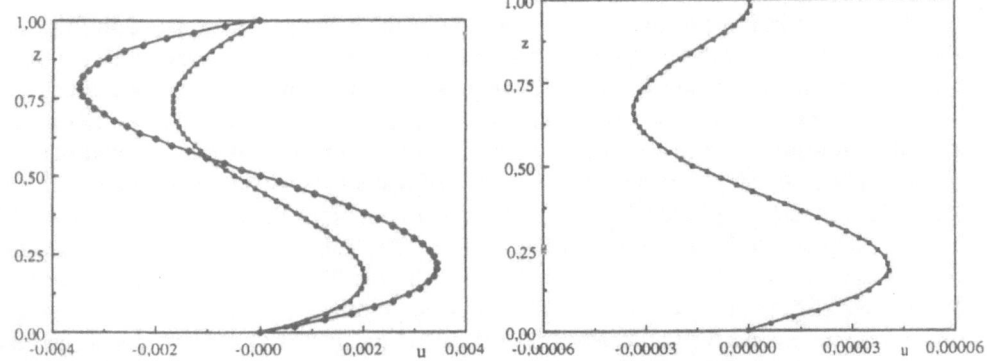

Figure 12. Axial variations of the radial velocity for DT3 cavity with 2.5 g at two radial locations. ◊: r = 0.5; ◆: r = 0.8.

Figure 13. Axial variations of the radial velocity for DT3 cavity with 3 g at radial location r = 0.98.

Contrary to the previous situation, a symmetry about the mid-plane z = 0.5 is observed on the flow structure when the acceleration is increased up to 20 g (Figs. 17a–d). At 0 g (Fig. 17a), the flow shows the typical thermal cell. Then, two counter-rotating cells, generated by increasing the centrifugal acceleration, develop symmetrically to the thermal cell, at the same locations as for cavity DT3 (Figs. 17b–d). This, finally, leads to a bending of the thermal cell for the highest acceleration, 20 g (Fig. 17d).

The axial variations of the radial component of the velocity are reported in Figs. 18a–b for two values of the centrifugal acceleration, 2.5 g and 20 g, respectively. They are displayed at radial locations r = 0.3 and 0.5. The velocity at the first location exhibits a

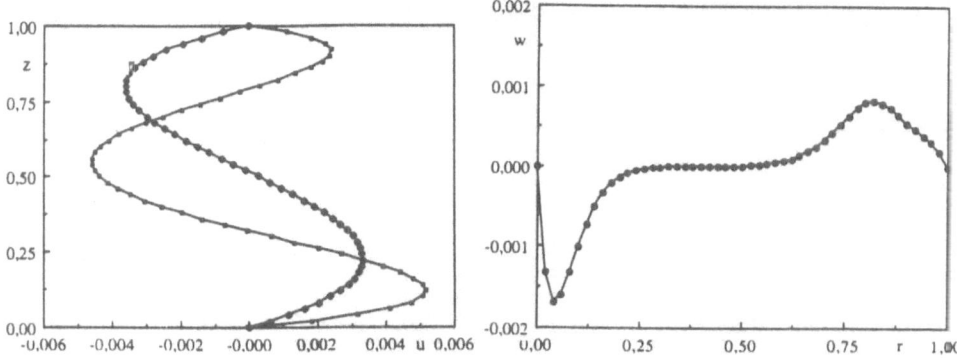

Figure 14. Axial variations of the radial velocity for DT3 cavity with 20 g at two radial locations. ◇: r = 0.5; ♦: r = 0.8.

Figure 15. Radial variations of the axial velocity for DT3 cavity with 20 g at axial location z = 0.5.

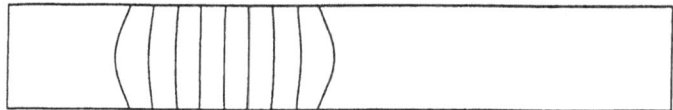

Figure 16. Isotherms for DT1 cavity at 0 g.

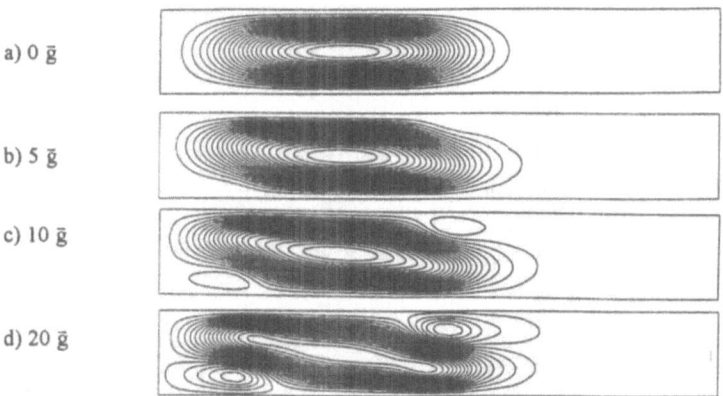

a) 0 ḡ

b) 5 ḡ

c) 10 ḡ

d) 20 ḡ

Figure 17. Flow configuration for DT1 cavity at different levels of centrifugal acceleration.

S-shaped profile for the two values of the centrifugal acceleration, characteristic of a buoyancy-driven regime. For the value 2.5 g, the fluid moves faster at location r = 0.3 than at r = 0.5, corresponding to the end of the thermal gradient and to the beginning of isothermal flow. For 20 g, the part corresponding to natural convection (z < 0.8) moves faster at r = 0.5 than at r = 0.3, in order to compensate the flow reversal generated by the centrifugal acceleration close to the upper wall.

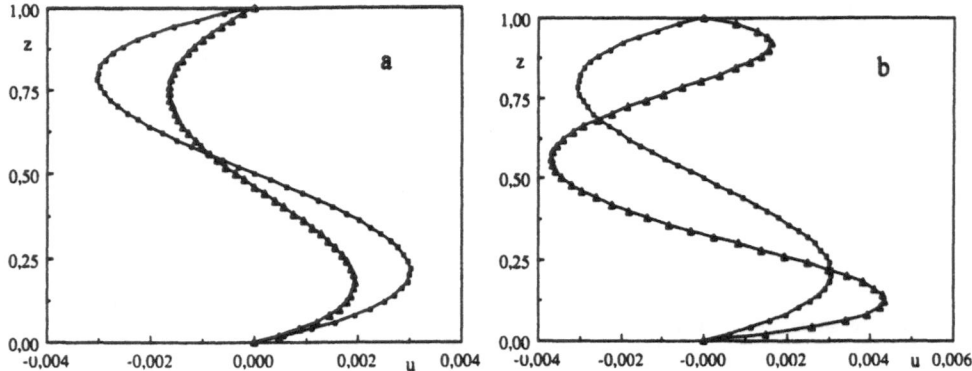

Figure 18. Axial variation of the radial velocity for DT1 cavity with 2.5 g (a) and 20 g (b) at two radial locations. ◆: r = 0.3; ■: r = 0.5.

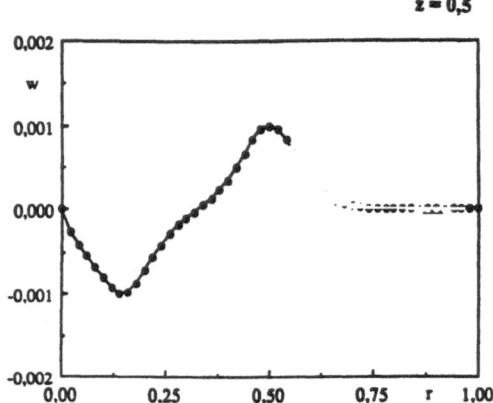

Figure 19. Radial variation of the axial velocity for DT1 cavity with 20 g at axial location z = 0.5.

The radial variation of the axial velocity is reported in Fig. 19 at axial location z = 0.5 for an acceleration of 20 g. According to the aforementioned flow structure, a symmetry is obtained for the buoyancy-driven regime part. After r = 0.625, the fluid behaves as a "solid." The same behavior is observed for all the values of the centrifugal acceleration considered, with different levels.

Analysis of the Results

A transition appears between the natural convection (BDR) and the centrifugal force dominated regimes (CER) below an acceleration of 5 g (about 3 g), for the DT1 and DT3 cavities. This centrifugal regime starts in the hot part of the ampoules. No significant change in the regime between 5 and 10 g was observed.

The symmetrical characteristic of the faceted seeds obtained at 5 g and 10 g is confirmed by the streamlines, which are developed in the source zone. The profiles of the velocity u = f(z) show this movement. The zone r < 0.2 is weakly affected by the increase of the acceleration level.

CONCLUSION

Our objectives were to measure the growth kinetics of GaAs on an oriented single crystal hemisphere in the presence of HCl, and to determine the influence of g on the growth behavior.

A thermodynamic study allowed us to determine the partial pressure of the different gaseous species in equilibrium with the solid phase GaAs and the theoretical flux in the cavity. The experimental fluxes were lower than predicted. Therefore, it will be necessary to take into account thermosolutal convection.

We have identified the crystallographic orientation of each facet of the epitaxial layers by analyzing the half spheres by Läue X-ray back diffraction.

The hydrodynamic model corresponded to the experimental conditions and exhibited axisymmetrical centrifugal behavior. The transition between the natural convection and the centrifugal regime takes place below 3.5 g for undoped GaAs and at a higher value for Sn-doped GaAs.

The axisymmetrical flow in the SAER furnace confirms the stabilization by the Coriolis effect with $F_c = 2V \times \Omega$ (V and Ω are the velocity and rotation rate, respectively).

Other experiments at g = 1 and under hypergravity must be considered, in order to determine the transitions between buoyancy-driven convection/Coriolis regime/centrifugal regime as a function of the Reynolds number, the temperature, and the partial pressure of GaCl.

A numerical simulation of three-dimensional flow regimes in a full cylinder is in progress, including a study of the effect of the orientation of the thermal gradient.

Acknowledgments

Support of parts of this research by CNES and by Aerospatiale is gratefully acknowledged.

We are also grateful to Dr. M. Laügt for meticulous preparation of the substrates and V. Oudomsak for skilled experimental advice and help. The authors wish to thank R. L. Sani for fruitful discussions, and C. Schiller for his useful advice on Läue diffraction.

The computations were performed on a CRAY-YMP2E with support provided by the Centre de Calcul de l'IMT (Château-Gombert, Marseille) and by the Conseil Regional PACA.

REFERENCES

1. D. Shaw, Kinetics aspects in the vapour phase epitaxy of III-V compound, *J. Cryst. Growth* 31:130 (1975).
2. R. Cadoret, Application of the theory of rates processes in the CVD of GaAs, *Mater. Sci.* 5:221 (1980).

3. A. Boucher and H. Hollan, Thermodynamics and experimental aspects of gallium arsenide vapor growth, *J. Electrochem. Soc.* 117:932 (1970).
4. R. Cadoret and M. Cadoret, A theoretical treatment of GaAs growth by vapour phase transport for {001} orientation, *J. Cryst. Growth* 31:142 (1975).

5. THERMODATA Data Bank, B.P. 66, 38402 F. Saint-Martin d'Hères.

6. J.L. Gentner, Vapour phase growth of GaAs by the chloride process under reduced pressure, *Philips J. Res.* 38:37 (1983).

7. R.C. Reid, J.M. Prauznitz, and T.K. Sherwood. "The Properties of Gases and Liquids," 3rd edn., McGraw-Hill, New York (1977).

8. D. Gottlieb and S. Orszag. "Numerical Analysis of Spectral Methods, Theory, and Applications," CBMS-SIAM Publications, Philadelphia (1977).

9. C. Canuto, M.Y. Hussaini, A. Quarteroni, and T.A. Zang. "Spectral Methods in Fluid Dynamics," Springer Verlag, New York (1988).

10. G.M. Homsy and J.L. Hudson, Centrifugally-driven thermal convection in a rotating cylinder, *J. Fluid Mech.* 35(1):33 (1969).

11. G. De Vahl Davis, E. Leonardi, and J.A. Reizes, Convection in a rotating annular cavity, *in*: "Proceedings XIVth Intern. Center for Heat and Mass Transfer Symposium," Dubrovnik, Yugoslavia (1982).

12. A. Randriamampianina, P. Bontoux, and B. Roux, Ecoulements induits par la force gravifique dans une cavité cylindrique en rotation, *Intern. J. Heat Mass Transfer* 30:1275 (1987).

13. A. Randriamampianina, Etude de régimes d'écoulements induits par la force gravifique dans une cavité cylindrique en rotation, Doctorate Thesis, Université Aix-Marseille II (1984).

14. C.W. Gear. "Numerical Initial Value Problems in Ordinary Differential Equations," Prentice-Hall Series in Automatic Computation, G.E. Forsythe, ed., Prentice-Hall, Englewood Cliffs, New Jersey (1971).

15. G.D. Byrne and A.C. Hindmarsh, Stiff ODE solvers: a review of current and coming attractions, *J. Comp. Phys.* 70:1 (1987).

16. L.R. Petzold. "Automatic Selection of Methods for Solving Stiff and Non-stiff Systems of Ordinary Differential Equations," Sandia National Lab. Report SAND80-8230, Sandia National Laboratory (1980).

17. A.C. Hindmarsh, LSODE and LSODI: two new initial value ordinary differential equation solvers, *ACM-SIGNUM Newsletter* 15:10 (1990).

18. D.B. Haidvogel and T.A. Zang, The accurate solution of Poisson's equation by expansion in Chebyshev polynomials, *J. Comp. Phys.* 30:137 (1979).

19. C. Temperton, Fast mixed-radix real Fourier transforms, *J. Comp. Phys.* 52:340 (1979).

20. G.P. Extremet, P. Bontoux, and B. Roux, Effect of temperature gradient locally applied on a long horizontal cavity, *Heat & Fluid Flow* 8(1):26 (1987).

FIRE BEHAVIOR IN MACROGRAVITY

Jie Chen,[1] Jean-Michel Most,[1] Pierre Joulain[1] and Daniel Durox[2]

[1]Laboratoire de Chimie Physique de la Combustion
Université de Poitiers - CNRS
Domaine du Deffend, France
F-86550 Mignaloux Beauvoir
[2]Laboratoire d'Aérothermique du CNRS
4ter route des Gardes
F-92190 Meudon, France

ABSTRACT

An enclosed pool fire was simulated by injecting ethane through a 6.2 cm diameter porous plate. The diffusion flame was investigated at low, normal and high gravity to 12 g. From scaling relations based on Froude modelling, the experimental flame characteristics (height, fluctuation frequency and radiant fraction of the flame) were correlated. The results were used for validation of numerical models and scale analysis.

INTRODUCTION

The most relevant feature that separates combustion from other branches of fluid physics is the existence of large temperature gradients in the combustion flow. The high exothermicity of the chemical reaction leads to large density variations and consequently to strong buoyant flows.

The main objective of this work is a better knowledge of the effects of gravity on the behavior of a diffusion flame representative of the first steps of a fire. During a fire, flames developing above a liquid or solid fuel are characterized by a low initial momentum (initial velocity U of the degradation products of the material) and by low values of the Froude number (inertia-gravity forces ratio) Fr:

$$Fr = \frac{U^2}{gL} \qquad (1)$$

The flames are fully buoyancy-driven when the Grashof number (gravity-inertia forces ratio) becomes greater than 0.1.

$$Gr = \frac{\beta_0 Tg\rho^2 L^3}{\mu^2} \qquad (2)$$

Numerous experimental and theoretical studies have been carried out on this kind of flame. Among the most important are the temperature and velocity measurements by McCaffrey,[1] chemical species fields determination by Orloff et al.[2] and Smith et al.,[3] entrainment rate of air by Delichatsios,[4] flame oscillations by Zukoski et al.[5] and Weckman et al.,[6] large organized structures by Schonbucher et al.,[7] and numerical modelling by Annarumma et al.[8]

The effect of buoyancy on such diffusion flames can be studied either by changing gravity, ambient pressure, or the initial fuel gas velocity U, hence the system thermal input. These last two possibilities allow the study of gravity effects for a constant value of the Reynolds number. In the present work, the Froude model, used to extrapolate the results of laboratory scale fire to large fires, is generalized. The scale factor is introduced through both the pressure and the gravity parameters. Buoyancy influences air entrainment in the diffusion flame and intensifies the formation of the eddy structure. Gravity affects the turbulent micro and macro scales, and controls the reacting system.

The literature on fire flames in normal earth's gravity and ambient pressure (Cetegen,[9] Chen,[10] Davis[11]) show the great influence of buoyant forces on flame stabilization, air entrainment into the reactive zone, diffusion flame flickering (eddy structure formation) and turbulence. Most of the papers on the influence of gravity deal with low or microgravity conditions (Durox[12]), and only a few studies deal with a high gravity environment. Altenkirch[13] studied the height and lift off laminar gas jet diffusion flames on a small cylindrical burner at high gravity. More recently, Katta[14] numerically modeled the dynamic behavior of H_2-O_2 diffusion flames. For large scale surfaces, the solid (or liquid) burning rate is mainly controlled by radiative processes between the flame and the degraded surface, but is mainly a function of the convective and diffusive heat transfer for small surfaces.

The objective of this study is to obtain a better knowledge of gravity effects on the structure (large scale eddy formation), size (height, envelope, volume), stability (flickering frequency), and radiant fraction of small size buoyancy driven diffusion flames. Empirical relationships are proposed and discussed using scaling techniques based on Froude modelling. The scale modeling is based on the governing laws of physics. Quintiere[15] shows examples of correlations successfully developed for a wide range of fire phenomena in terms of the significant dimensionless groups for Froude modelling.

EXPERIMENTAL SET-UP

The horizontal surface burning of a solid polymer pool fire was partly simulated by injecting fuelled gas through a 6.2 cm diameter D water cooled porous plate. Ethane was selected as the closest component of standard polymer degradation products during a fire.

The experimental set-up is presented in Fig. 1. The burner to simulate a pool fire was enclosed in a 0.3 m diameter and 1.0 m height vessel to obtain steady conditions. The air needed for combustion was quietly injected in the lower part of the vessel. The vessel walls had three silicate windows to allow flame visualization and optical measurements.

Visualization of the spontaneous flame emission was made. The video images were numerically processed in order to determine the mean flame characteristics (contour, height,

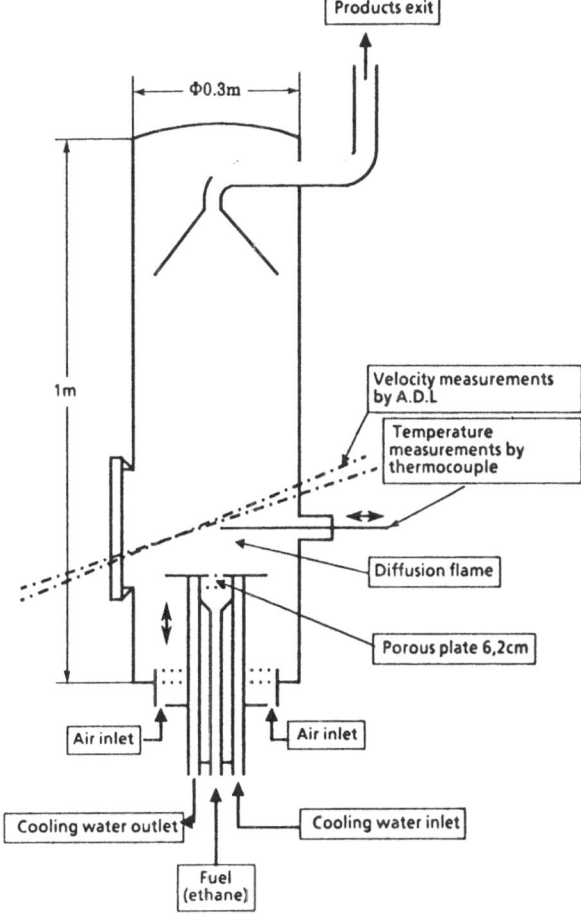

Figure 1. Experimental set-up of the simulated pool fire.

surface and volume) from 120 successive frames. The oscillation frequency was obtained by direct analog processing of the photo detector output signal on the video screen. The temperature fields were measured with a thin 75 μm diameter Cr/Al bare wire thermocouple. The mean temperature was calculated from 1536 instantaneous values. Stepping motors moved the burner axially up and down and the thermocouple radially. A wide angle radiometer (Medtherm radiometer with sapphire window located at distance R from the flame axis, R=0.22 m) measured the global radiative heat flux from the flame. The experiment (flame parameters, moving tables, data acquisition, etc.) was monitored from the computer terminals installed in the centrifuge control room.

The high g tests were done in a centrifuge facility. The whole experimental set-up (vessel, gas burner, ethane and air supplies and video recorder) was installed aboard the basket. Data acquisition and processing and the minicomputer were at the centrifuge axis. The four experimental conditions were 3, 6, 9 and 12 times earth's gravity g. For a hot products velocity of 1.5 m/s, the ratio of Coriolis to centrifugal accelerations was 4.26×10^{-2} at 3 g, and decreased as the g-level increased (1.5×10^{-5} at 6 g, etc.), and thus is neglected. Some experimental tests also were performed in reduced gravity conditions during parabolic aircraft flights.

We studied an ethane flame with 1.04 kW thermal output. The ethane flow rate was 1.1 l/min, corresponding to a mass flow rate of 7.52×10^{-3} kg/m²/s. The air mass flow rate was 60 l/min. The 1 g Froude number was 6×10^{-5}.

RESULTS

Physical Appearance of the Flames

The flames examined at normal earth's gravity (reference condition) showed three main zones.[1] Close to the burner surface edge, the flame presented a luminous annular zone around a fuel-rich central region. Figure 2 shows the physical appearance of the flame from microgravity to 12 g. Most of the chemical reactions occur in the luminous zone, due to the availability of oxidizer brought by the entrainment of air. The large heat release induces a drastic increase in the gas velocity and temperature. The flame appears nearly laminar with a light blue color. Downstream, the flame (third zone) turns yellow as the temperature maximum reaches the flame axis.

By increasing gravity, the flame behavior is visibly modified. First, the flame changes towards the blue and is more luminous (at 12 g the whole flame is blue). Its height decreases. These phenomena are interpreted as an increase of the combustion efficiency with a more complete chemical reaction (fuel radicals emission). The air entrainment is faster and soot formation is weaker. Also the flame becomes compact, its neck is less visible, its height decreases, and the reaction zone moves toward the burner surface. Finally, flame oscillation increases.

At 0.2 g the flame is compact, cylindrical, yellow and stable. During the microgravity period, a flat flame seems to float over the burner, and any small force perturbs the flow.

Flame Height

The flame height L_f corresponds to the boundary of visible gas emission and is related either to the end of soot oxidation or to burning of the hydrocarbon radicals. The value of L_f was obtained from the processing of 120 flame frames and is defined by the 50% flame appearance rate.

Figure 3 shows that the flame height decreased with $g^{-1/3}$ from 0.2 g to 12 g.

$$L_f \propto g^{-1/3} \tag{3}$$

These results are well correlated by using the dimensional analysis of Quintiere,[15] as shown in Eq. (4), and in agreement with the experimental correlation by Zukoski et al.[5] for fire plumes with $Q^* < 1$.

$$L_f/D \propto Q^{*2/3} \tag{4}$$

where

$$Q^* = \frac{Q}{\rho_\infty C_{p\infty} T_\infty \sqrt{g} D^{5/2}} \tag{5}$$

is the fire Froude Number, and ρ_∞ , $C_{p\infty}$, and T_∞ are, respectively, density, heat capacity and temperature.

These results show that gravity influences the flame diameter in the lower reactive zone. At high gravity, the reduction of the flame diameter is due to buoyancy forces, which tend to decelerate unburned cold fuel gases whose density is greater than the ambient density and accelerate those of smaller density, as interpreted by Altenkirch et al.[13]

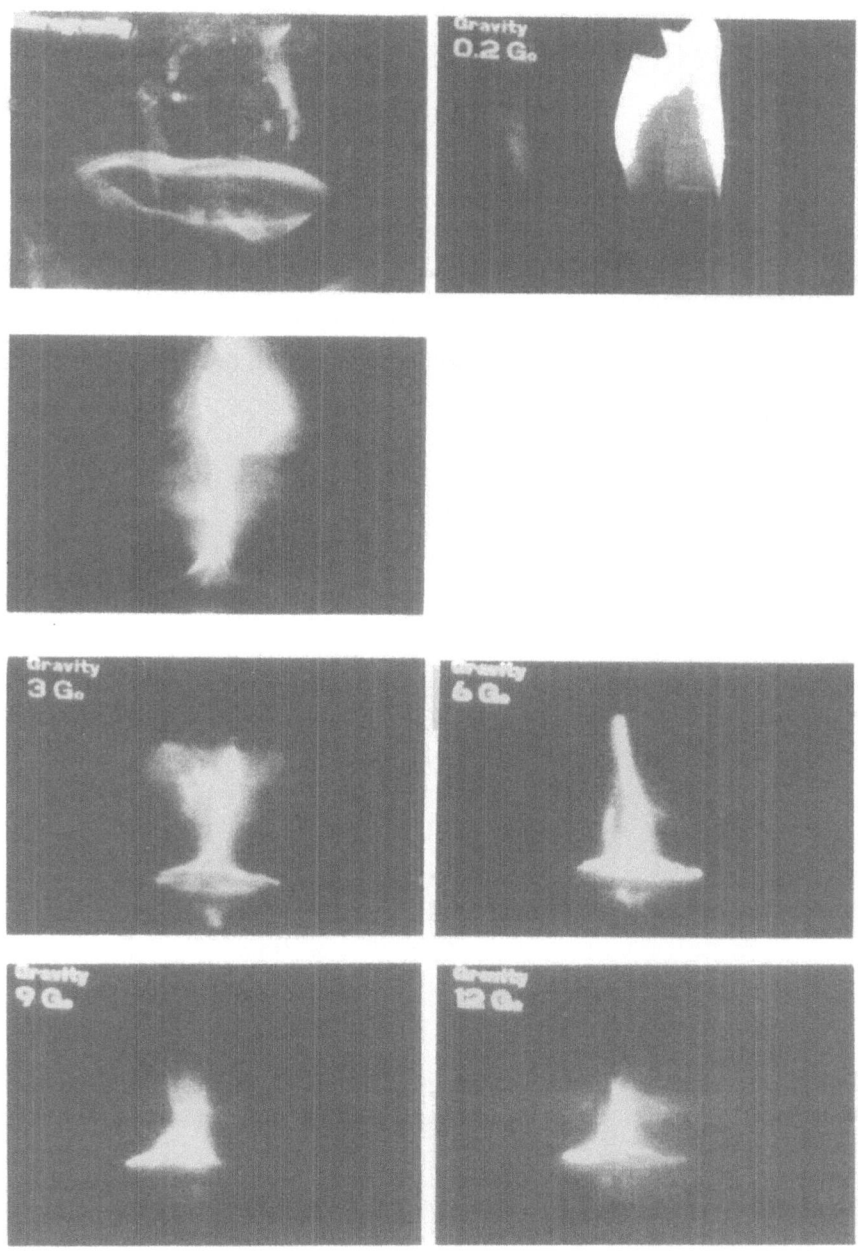

Figure 2. Influence of gravity forces on a simulated 6.2 cm diameter pool fire. Reduced gravity: microgravity 0.2 g. Normal earth's gravity: 1 g. High gravity: 3 g, 6 g, 9 g, 12 g.

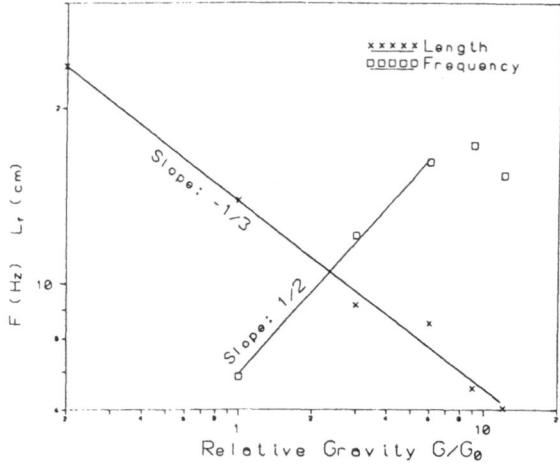

Figure 3. Influence of gravity on the height and oscillation frequency of the flame for a 1 kw ethane simulated pool fire.

Flame Fluctuation

The flame fluctuation frequency is obtained from fast Fourier transformation of the analogue output signal of a photodetector placed on the monitor screen. For buoyant flow, the fluctuation frequency can be expressed as:

$$f = u/L \qquad (6)$$

where

$$u = \sqrt{2\left(\frac{\rho_0}{\rho} - 1\right)zg + u_0^2} \qquad (7)$$

Here u, L are the characteristic velocity and length. The initial velocity u_0 is negligible for a buoyant flame. The length scale is proportional to the burner diameter. Thus Eq. (7) can be written:

$$f \propto \sqrt{g/D} \qquad (8)$$

For normal gravity, the results of numerous experimental studies for various fuels are well correlated by Eq. (8). Our experimental results (Fig. 3) are in agreement with Eq. (8). The frequency decreases at high g only due to a video sweeping that is too slow (25 frames/s), leading to biased measurements.

Radiative heat flux

The total radiant power X_{RAD} is computed from the measured irradiance H using the Medtherm radiometer assuming spherical emission. The radiant fraction X_{RAD} is a constant for a given gas and is equal to 0.21 for ethane.[16] The radiant output from the radiating gas volume can be expressed as:

$$Q_{RAD} = 4\pi S^2 H \tag{9}$$

$$X_{RAD} = \frac{\sigma T_f^4[1 - e^{-KL}]A_f}{Q} \tag{10}$$

where $L = 3.6\, V_f/A_f$ is the radiation path length, T_f the effective radiation temperature, and K the absorption coefficient. K is related to the soot concentration Y_s and depends on the Kolmogorov time t_s and on the chemical time t_{ch}.[17] When the acceleration is greater than 1 g, Y_s is low (blue flame), and T_f and K stay practically constant. So X_{RAD} is only a function of the envelope area (L staying constant with g). Fig. 4 shows X_{RAD} versus g, corresponding to $X_{RAD} \propto g^{-0.3}$. This result is related to a decreasing measured flame surface with increasing acceleration by $A_f \propto g^{-0.33}$.

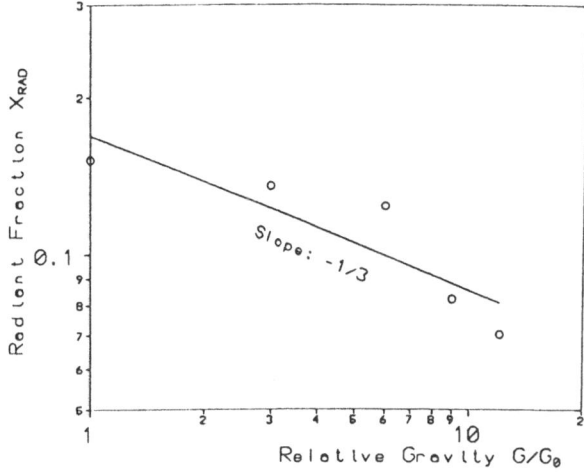

Figure 4. Influence of gravity on the radiant fraction X_{RAD} for a 1 kW ethane simulated pool fire.

Temperature Fields

The axial temperature evolution (normalized by the mean flame height; L_{f0} is the flame height for 1 g) is presented in Fig. 5 for various gravity levels. When acceleration increases, the maximum temperature decreases (as a function of $g^{-1/3}$) and its location moves toward the burner. It is very difficult to correlate this result with the flame observation.

The blue color at high gravity indicates a low soot concentration. Then the heat released in the flame should be higher.

Increasing acceleration may increase air entrainment, modifying the local stoichiometry and lowering the flame temperature. At 3 g, Coriolis acceleration perturbs the flame and the temperature profiles.

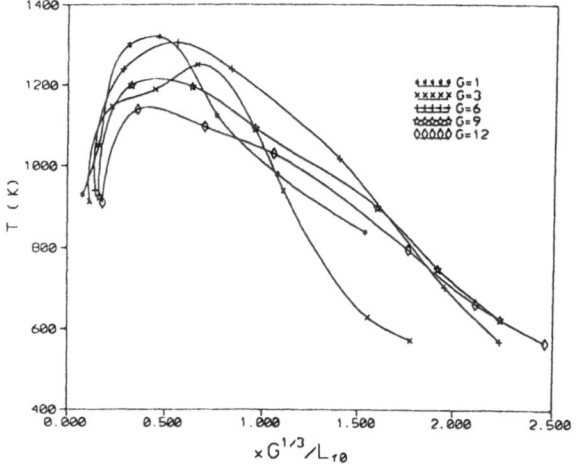

Figure 5. Axial mean temperature evolution in the flame normalised by the flame height $(g^{-1/3}/L_{f0})$. L_{f0} is the flame height for 1 g.

CONCLUSION

A laboratory scale, simulated pool fire (6.2 cm diameter) was experimentally studied at high gravity. Buoyancy effects were evaluated for the height, frequency fluctuation and radiance of the diffusion flame from 1 to 12 g. The main results show:

-The experimental flame height L_f is correlated with the fire Froude number when the length scale used is the flame diameter close to the burner surface ($L_f \propto g^{-1/4}$).

-Buoyancy affects the flame fluctuation frequency f through an aerodynamic effect ($f \propto g^{-1/2}$).

-The radiant fraction is mainly a function of the flame volume at high gravity ($X_{RAD} \propto g^{-0.3}$), and on the soot concentration through the absorption coefficient and the flame area.

ACKNOWLEDGEMENTS

The authors gratefully acknowledge the Laboratoire Central des Ponts et Chaussées, and particularly Mr. Garnier, for help and technical assistance during our experiments in their centrifuge facility. The contribution of Mr. B. Sztal and J. Baillargeat is acknowledged. This study was supported by CNRS, CNES, ESA (ESA parabolic flight campaign in January 1993) and the Ministère de la Recherche et de la Technologie (Comité de Valorisation de la Centrifugeuse in France). We thank these agencies and ministries.

REFERENCES

1. B.J. Mc Caffrey, "Purely buoyant diffusion flames: some experimental results," NBSIR 79-1910, National Bureau of Standards, Washington, D.C. (1979).
2. L. Orloff, J. de Ris and M.A. Delichatsios, General correlations of chemical species in turbulent fires, *in:* "Proceedings of the Twenty-first International Symposium on Combustion," The Combustion Institute, (1986) p101.
3. D.A. Smith and G. Cox, Major chemical species in buoyant turbulent diffusion flames, *Combustion and Flame*, 91:226 (1992).
4. M.A. Delichatsios, Air Entrainment into buoyant jet flames and pool fires, *Combustion and Flame*, 70:33 (1987).
5. E.E. Zukoski, B.M. Cetegen and T. Kubota, Visible structure of buoyant diffusion flames, *in:* "Proceedings of the Twentieth International Symposium on Combustion," The Combustion Institute (1984) p361.
6. E.J. Weckman and A. Sobiesak, The oscillatory behavior of medium-scale pool fires, *in:* "Proceedings of the Twenty-Second International Symposium on Combustion," The Combustion Institute, (1989) p1299.
7. A. Schonbucher, B. Arnold, V. Barnhardt, V. Bieller, H. Kasper, M. Aufmann, R. Lucas, and N. Schiess, Simultaneous observation of organized density structures and the visible field in pool fires, *in:* "Proceedings of the Twenty-third International Symposium on Combustion", The Combustion Institute (1986) p83.
8. M. Annarumma, J.M. Most and P. Joulain, On the numerical modelling of buoyancy-dominated turbulent vertical diffusion flames, *Combustion and Flame*, 85:403 (1991).
9. B.M. Cetegen and T. Ahmed, Experiments on the periodic instability of buoyant plumes and pool fires, *Combustion and Flames* (submitted).
10. L.D. Chen and J.P. Seabe, Buoyant diffusion flames, *in:* "Proceedings of the Twenty-second International Symposium on Combustion," The Combustion Institute (1988) p677.
11. R.W. Davis *et al.*, Preliminary results of numerical-experimental study of the dynamic structure of a buoyant jet diffusion flame, *Combustion and Flame*, 83:263 (1990).
12. D. Durox *et al.*, Some effects of gravity on the behavior of premixed flames, *Combustion and Flame*, 82:66 (1990).
13. R.A. Altenkirch, R. Eichorn, N.N. Hsu, A.B. Brancic and N.E. Cevallos, Characteristics of laminar gas jet diffusion flames under the influence of elevated gravity, *in:* "Proceedings of the Sixteenth International Symposium on Combustion," The Combustion Institute (1976) p1165.
14. V.R. Katta, I.P. Goss and W.M. Roquemore, Numerical investigations on the dynamic behavior of a H_2-N_2 diffusion flame under the influence of gravitational force, AIAA Paper 92-0335 (1992).
15. J.G. Quintiere, Scaling applications in fire research, *Fire Safety Journal*, 15:3 (1989).
16. G.M Markstein, Relationship between smoke point and radiant emission from buoyancy turbulent and laminar diffusion flames, *in:* "Proceedings of the Twentieth International Symposium on Combustion," The Combustion Institute (1984) p1055.
17. M.A. Delichatsios and L. Orloff, Effects of turbulence on flame radiation from diffusion flames, *in:* "Proceedings of the Twenty-Second International Symposium on Combustion," The Combustion Institute (1993) p1271.

CORIOLIS EFFECT ON HEAT TRANSFER
EXPERIMENT USING HOT-WIRE TECHNIQUE
ON CENTRIFUGE

Taketoshi Hibiya,[1]* Shin Nakamura,[1]*, Kyung-Woo Yi,[2] and
Koichi Kakimoto[2]

[1]Space Technology Corporation
 34 Miyukigaoka
 Tsukuba 305, Japan
[2]Fundamental Research Laboratories
 NEC Corporation
 34 Miyukigaoka
 Tsukuba 305, Japan

ABSTRACT

A transient hot-wire technique was applied to examine the influence of the Coriolis effect on heat transfer on a centrifuge. A thermal conductivity measurement facility, once flown on board the TEXUS-24 rocket, was set on the 7.25 m rotating arm of the centrifuge. The temperature increase of the sensing wire, which was on a solid state substrate immersed in mercury, depended not only on input power and rotational acceleration but also on the orientation of the specimen. The temperature increase was affected by the Coriolis force, depending on the orientation: enhancement or suppression of heat transfer from the wire by convection.

INTRODUCTION

The thermal conductivity of molten InSb was successfully measured under microgravity using the thermal conductivity measurement facility (TCMF) on board the German souncing rocket TEXUS-24, and in a newly constructed drop shaft at the Japan Microgravity Center (JAMIC) in Hokkaido, Japan.[1-3] Through this experiment, it was proved that microgravity is very effective in applying a transient hot-wire technique to measurement of the thermal conductivity of liquids. The influence of convection on the measurement has already been

---

*On leave from NEC Corporation.

Materials Processing in High Gravity, Edited by L.L. Regel
and W.R. Wilcox, Plenum Press, New York, 1994

discussed in detail.[2,4] Convection both before and during the measurement was considered. The microgravity level required for the measurement has also been made clear through a parabolic flight experiment, which supplies a low gravity level of 10^{-1} and 10^{-2} g.[4] Through this experiment, gravitational acceleration was recognized as a continuously variable materials processing parameter.

On the other hand, the effect of high gravitational acceleration, more than 1 g, on the measurement is not yet clear, although a reference measurement during the launch of the TEXUS-24 rocket under 10 g condition has been carried out.[1] A centrifuge is the only possible way to generate high gravitational acceleration on earth, but the effects due to rotation must be considered: the effects of Coriolis and centrifugal forces. Müller's group[5,6] and Rodot et al.[7] reported crystal growth experiments and discussed the changes in flow mode in a crystal growth cell that has a temperature gradient. Weber et al.[6] reported that a flow mode change took place due to the Coriolis effect. Ramachandran et al.[8] numerically modeled heat and mass transfer for a cell having a temperature gradient along the rotating arm. Until now, there have been no reports on a hot-wire technique using a centrifuge, where the line heat source is set parallel to the acceleration.

This paper discusses the orientation dependence of the Coriolis effect on the temperature increase for a sensing wire fabricated on a solid state substrate, when a transient hot-wire technique is used on a centrifuge.

2. EXPERIMENT

A transient hot-wire technique with a ceramic probe was employed to examine the temperature increase of a sensing wire immersed in mercury on a centrifuge. Mercury was selected as a model fluid because it has a small Prandtl number ($Pr = 0.026$).

If there is no convective heat transfer, the temperature increase for the wire on the substrate is:[9]

$$\delta T = [Q/2(\lambda_L + \lambda_S)] \ln t + C \qquad (1)$$

where λ_L and λ_S are the thermal conductivities of mercury and of the alumina substrate, respectively. Here Q is the input electric power per unit length of wire, t is time, and C is a constant.

The flight model of the TCMF, flown once on board the TEXUS-24 rocket, was set on the 7.25 m rotating arm of the centrifuge at the Tsukuba Space Center for the National Space Development Agency of Japan (NASDA), as shown in Fig. 1. The specimen consisted of mercury and a 100-mm long solid state substrate installed within a carbon crucible. On the surface of the substrate a 70-mm long sensing wire was fabricated using a printing technique.[1] Centrifugal acceleration was changed (2.0, 5.0 and 10.0 g) by changing the rotation rate of the arm (1.64, 2.60 and 3.68 rad/s, respectively). The stability of centrifugal acceleration was within ± 5%. The angle of the TCMF was changed using a wedge that depended on the rotation rate of the arm, so that the sensing wire was always parallel to the vector sum of the gravitational and centrifugal accelerations, as shown in Fig. 2. In order to examine how convective heat transfer affects the temperature increase of the sensing wire, the orientation of the specimen was changed, as shown in A, B, C, and D of Fig. 3. Figure 3 shows the orientation as observed from the bottom of the crucible. As a reference, the measurement was also carried out under 1 g in orientation X, where the sensing wire was set horizontally on the surface of mercury, so that convection due to heating the wire was suppressed to a great extent. The input electric power was 273, 600 and 759 W/m, corresponding to input currents of 1.5, 2.0 and 2.5 A, respectively.

Figure 1. Centrifuge of the Tsukuba Space Center, the National Space Development Agency of Japan (NASDA).

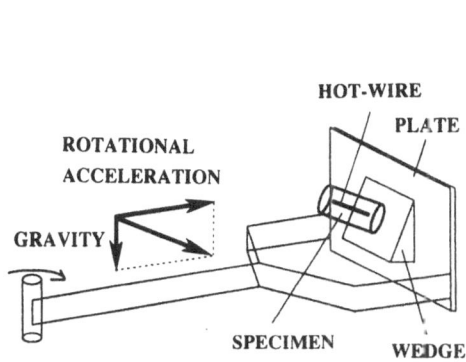

Figure 2. Geometry of the experiment.

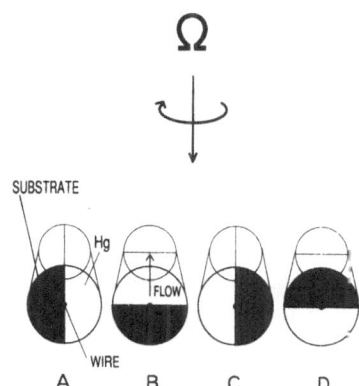

Figure 3. Orientation of the specimens as observed from the bottom of the crucible.

3. TEMPERATURE INCREASE OF THE HOT–WIRE ON THE CENTRIFUGE

Figures 4a, b, c, d and e show the temperature increase of the sensing wire immersed in mercury on the centrifuge. In these figures is plotted the temperature increase compared to the temperature 1.1 s after starting the current. For an input current of 1.5 A and Orientation X, the temperature of the wire at 1.1 s after the start of the current had already increased by 9.25 K from room temperature: for 2.0 A it had increased by 18.4 K and for 2.5 A by 30.5 K. The temperature increase of the sensing wire depended on the input heat power, acceleration level, and orientation of the specimen. The temperature increase curves were no longer smooth, but showed oscillation. The number and amplitude of the oscillation became more remarkable with increasing acceleration level and input power. Oscillation were also observed for the temperature increase curve for a wire immersed in molten InSb, during ignition of the second stage engine of the TEXUS–24 rocket, when an acceleration level of 10 g was recorded.[2]

The temperature increase of the wire was suppressed by increasing the acceleration. This was observed as a deviation from the temperature increase curve for Orientation X (horizontal under 1 g). The temperature increase in the initial stage, i.e., up to 3 s after turning the power on, depended significantly on the specimen orientation. The order of the temperature increase was as follows:

$$\text{Orientation } A > B = D > C.$$

Orientations B and D showed almost the same deviation from a linear relationship. For all orientations, the larger the acceleration level and input electric power, the earlier the deviation from X began.

A crossing of the temperature increase curves for Orientations A and B was observed, e.g., 3 s after power on at 10 g with a current of 2 A.

Figures 5a, b, c and d show the apparent thermal conductivity obtained from the experimental results shown in Figs. 4a, b, c and d, using Eq. (1). For the measurement at 10 g, an apparent thermal conductivity could not be plotted because of the strong oscillation of the data. As shown in Figs. 5a, b, c and d, the apparent thermal conductivity was larger for Orientation C at the beginning of the measurement, e.g., up to 3 s for 5 g and 2.5 A.

In order to see the effect of the specimen configuration and the Coriolis force on heat transfer from the sensing wire to mercury, a parameter ζ was introduced, as follows:

$$\zeta = (\lambda_L + \lambda_S + \lambda_C) / (\lambda_L + \lambda_S) \ . \tag{2}$$

The parameter ζ is similar to the Nusselt number, Nu, for convective heat transfer, and is the ratio of total heat transfer to conductive heat transfer.[4] Figure 6 shows ζ as a function of the Rayleigh number, Ra, for the present measurement cell. The parameter ζ was calculated using the apparent thermal conductivities, $\lambda_L + \lambda_S + \lambda_C$, obtained 3.0 s after turning on the current. Here λ_C is the contribution of convective heat transfer to the apparent thermal conductivity. For $\lambda_L + \lambda_S$, we used the apparent thermal conductivity 1.1 s after starting the current for Orientation X. The Rayleigh number is defined as follows:

$$Ra = g \, \Delta T \, \beta \, L^3 / \kappa \, v \tag{3}$$

where ΔT is the temperature difference between the sensing wire and the wall. In calculating the Rayleigh number for the centrifuge experiments, the total acceleration was used. For convenience, ΔT was taken as the temperature increase of the sensing wire 3.0 s after the start of current. Here, L is distance between the sensing wire and the inner wall of the crucible (14 mm), β is the volumetric expansion coefficient (1.82×10^{-4}/K), κ the thermal diffusivity (4.44×10^{-6} m^2/s), and v the kinematic viscosity (1.15×10^{-7} m^2/s) of

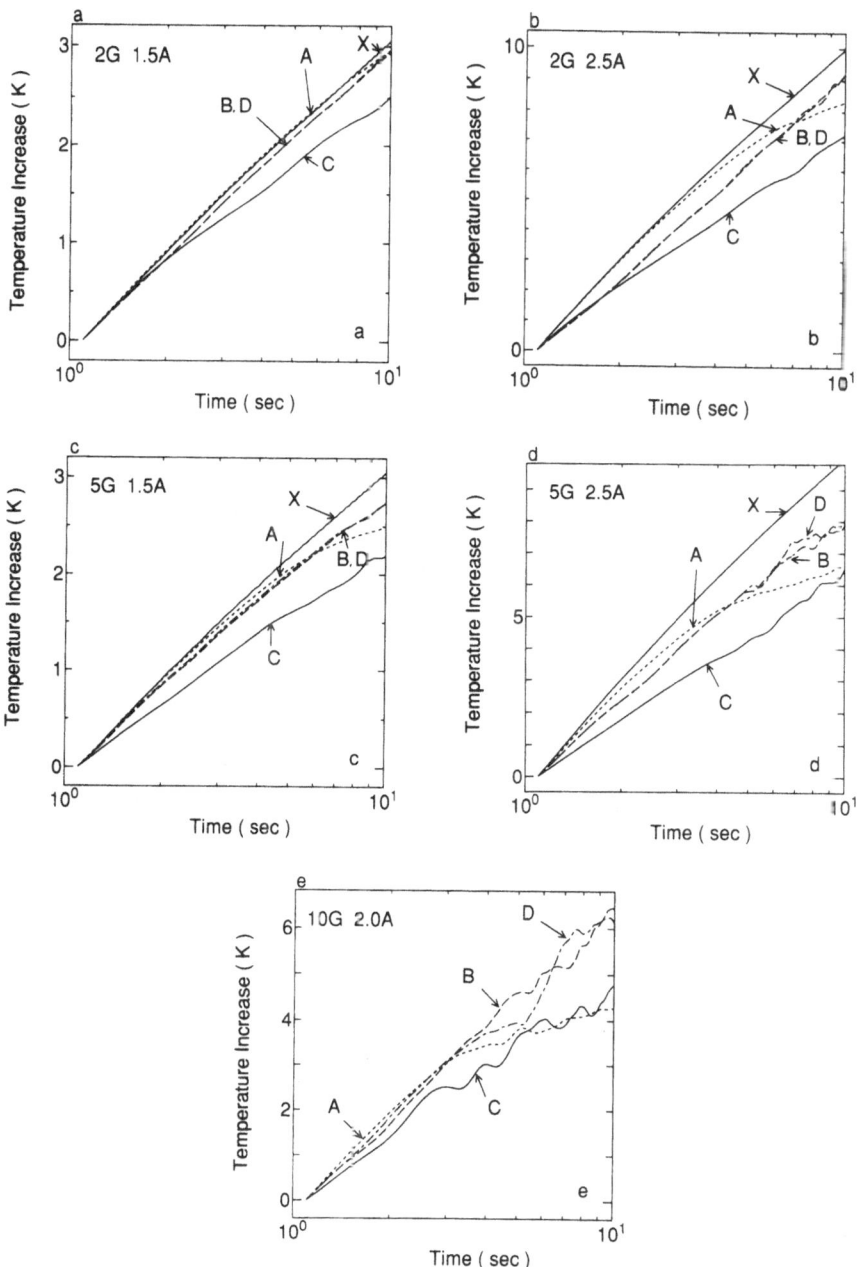

Figure 4. Temperature increase curves for different combinations of rotational acceleration and input electric current: (a) 2 g and 1.5 A, (b) 2 g and 2.5 A, (c) 5 g and 1.5 A, (d) 5 g and 2.5 A, and (e) 10 g and 2.0 A.

mercury. In Fig. 6 is plotted the value of ζ calculated from the thermal conductivity measurements on mercury under low gravity using a parabolic flight of the aircraft and on earth. All of these data were obtained 3 s after the start of the current.

Figure 6 shows that ζ values obtained from the centrifuge experiments were not on the line extrapolated to the corresponding Rayleigh numbers from the low- and earth-gravity

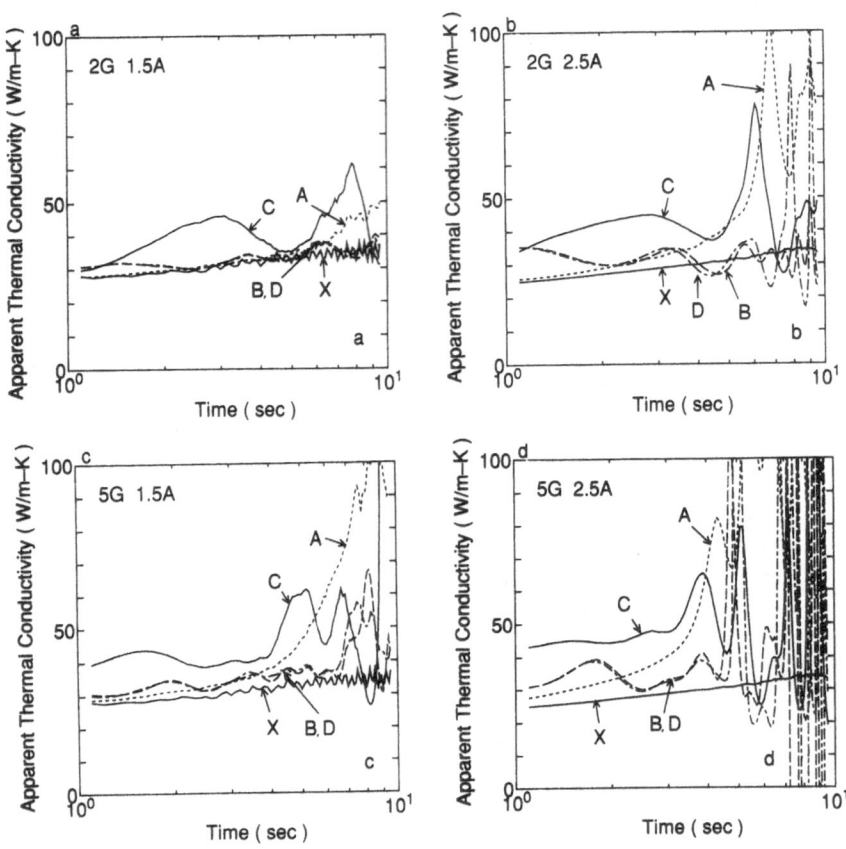

Figure 5. Apparent thermal conductivity for different combinations of rotational acceleration and input electric current: (a) 2 g and 1.5 A, (b) 2 g and 2.5 A, (c) 5 g and 1.5 A, and (d) 5 g and 2.5 A.

experiments. The value of ζ was also dependent on the orientation of the specimen. The ζ value for Orientation C was above the extrapolated line, while ζ was smaller for Orientations A, B and D. The ζ values for Orientation C were scattered. This is due to the oscillations in the apparent thermal conductivity shown in Fig. 5. Deduced ζ values for Orientation C are depicted by a dashed line.

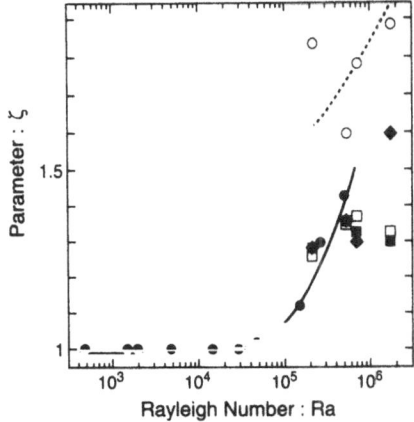

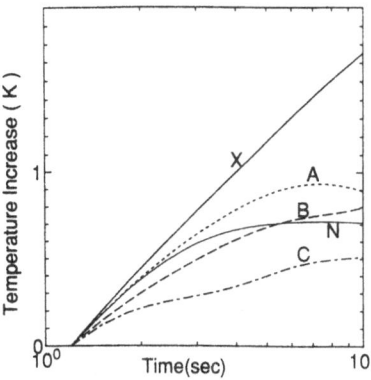

Figure 6. Parameter ζ which shows the ratio of total heat transfer to conductive heat transfer, as a function of the Rayleigh number: ● obtained under low- and earth-gravity, ♦ Orientation A, ■ Orientation B, ○ Orientation C, and □ Orientation D.

Figure 7. Calculated temperataure increase for rotational acceleration of 5 g. Orientation X: conduction only. N: the Coriolis effect ignored. For Orientations A, B and C, see Fig. 3.

4. DISCUSSION

The experimental results obtained on the centrifuge, particularly their dependence on orientation, can be interpreted by the Navier–Stokes equation for a rotating fluid:

$$(\partial/\partial t + u \cdot \nabla)u + 2\,\Omega \times u + \Omega \times (\Omega \times r) = -\nabla p/\rho - g + F/\rho \qquad (4)$$

where t is time, u is the axial (parallel to the rotating arm) velocity of the fluid, Ω is rotation rate, r is distance from the axis of the centrifuge, p is pressure, ρ is the density of the liquid, and F is the body force. The second term in the left–hand side of the equation shows the Coriolis effect. Depending on the vector of the axial flow, i.e., inbound or outbound, acceleration or deceleration of the radial flow takes place. In the present measurement cell, the flow is generated by heating the wire and the direction is inbound (Figs. 2 and 3).

A three–dimensional calculation was carried out to explain the experimental results.[10,11] Figure 7 shows the calculated temperature increases for several orientations with an input current of 1.5 A and a centrifugal acceleration of 5 g. For Orientation N, the calculation was carried out ignoring the Coriolis effect. The calculated temperature increase is similar to the experimental result. The order of the temperature increase was the same up to 3 s after the start of current: Orientation A > B = B > C. The order of the temperature increase was explained by heat transfer from the wire to the mercury and the substrate by buoyancy–driven convection modified by the Coriolis effect depending on the Orientation.[10,11] In Orientation A the temperature increase was the largest, because the Coriolis force pushes the radial flow initiated by wire heating to the solid substrate and suppresses heat transfer by the radial flow. On the other hand, for Orientation C, the

Coriolis force peels the flow and enhances a radial flow, which transfers heat from the wire into the mercury. For Orientations B and D, the Coriolis force has the same effect; it develops a radial flow parallel to the substrate surface. Because the characteristic length of the crucible for B, which is defined parallel to the Coriolis force, is larger than for A and C, the flow velocity is also larger than for A and C.[10,11] The radial flow not only transfers heat from the hot-wire to the mercury, but also enhances heat transfer from the substrate to the mercury and then transfers it to the vicinity of the wire, so that the temperature of the wire can be kept high. Therefore, the temperature increase for B is larger than for C, but smaller than for A, even though the calculated radial velocity is larger for B than for C.

As shown in Figs. 4a, b, c, d and e, the temperature increase curves showed oscillations. Oscillations were also produced by the numerical calculations. This means that the flow becomes oscillatory with an increase in centrifugal acceleration and input power, depending on the configuration. Crossing of the temperature increase curves for Orientations A and B was observed in the experiments. The larger was the centrifugal acceleration, the earlier was the crossing. Although the calculation did not show a crossing by 10 s after the start of current, the curves for A and B seemed to cross after 10 s. The reason for crossing of the temperature increase curves is not known.

The contribution of convection to heat transfer shown in Fig. 6 can be explained to some extent by the numerical simulation. As shown in Fig. 7 for acceleration without the Coriolis effect, the temperature increase of the wire is larger for Orientation N than for C. This supports the experimental result shown in Fig. 6, in which convective heat transfer from the wire for C was more pronounced than expected by extrapolation from the low and earth-gravity experiments. Also, the trend of ζ values for A is supported by numerical calculation. However, the smaller ζ values for B and D than predicted by numerical calculation are not understood, particularly at higher Rayleigh number; although the existence of oscillations in the apparent thermal conductivity curve was taken into account to explain the small ζ values for 2.5 A and 5.0 g. According to the calculated results shown in Fig. 7, ζ for B and D is expected to be larger than for N and smaller than for C.

CONCLUSION

A hot-wire technique was used on a centrifuge. Mercury was used as a model fluid with a low Prandtl number. The temperature increase of the wire was dependent on the orientation of the specimen, because the Coriolis force modified the convective heat transfer from the wire. The magnitude of heat transfer on the centrifuge was different from that extrapolated from low g and 1 g results. This is because the Coriolis force either increases or decreases heat transfer depending on the orientation.

Acknowledgments

The authors would like to express their thanks to M. Eguchi, M. Watanabe, T. Yoshiura, T. Yokota and F. Yamamoto of NEC Corporation for preparing the centrifuge experiment. Thanks are also due to I. Otsu, M. Saito and N. Ogura of the National Space Development Agency of Japan (NASDA) for this support in carrying out the experiments on the centrifuge in the Tsukuba Space Center of NASDA.

REFERENCES

1. S. Nakamura, T. Hibiya, F. Yamamoto, and T. Yokota, Measurement of the thermal conductivity of molten InSb under microgravity, *Thermophys.* 12:783 (1991).

2. S. Nakamura and T. Hibiya, Measurement of the thermal conductivity of molten InSb in a drop shaft, *in*: "Proceedings of the 8th European Symposium on Materials and Fluid Sciences under Microgravity," Brussels (1992), p. 233.

3. F. Yamamoto, S. Nakamura, T. Hibiya, T. Yokota, D. Grothe, H. Harms, and P. Kyr, Developing a measuring system for thermal conductivity using transient hot-wire method under microgravity, *in*: "Proceedings of the CSME Mechanical Engineering Forum," Toronto (1990), p. 1.

4. S. Nakamura, T. Hibiya, and F. Yamamoto, Effect of convective heat transfer on thermal conductivity measurements under microgravity using a transient hot-wire method, *Microgravity Sci. & Technol.* 5:156 (1992).

5. G. Müller, E. Schmidt, and P. Kyr, Investigation of convection in melts and crystal growth under large inertial acceleration, *J. Cryst. Growth* 49:387 (1980).

6. W. Weber, G. Neumann, and G. Müller, Stabilizing influence of the Coriolis force during melt growth on a centrifuge, *J. Cryst. Growth* 100:145 (1990).

7. H. Rodot, L.L. Regel, G.V. Sarafanov, M. Hamidi, I.V. Videskii, and A.M. Turtchaninov, Cristaux de tellurure de plomb elaborés en centrifugeuse, *J. Cryst. Growth* 79:77 (1986).

8. N. Ramachandran, J.P. Downey, P.A. Curreri, and J.C. Jones, Numerical modeling of crystal growth on a centrifuge for unstable natural convection configurations, *J. Cryst. Growth* 126:655 (1993).

9. E. Takegoshi, S. Imura, Y. Hirasawa, and T. Takeda, A method of measuring the thermal conductivity of solid materials by transient hot-wire method of comparison, *Bull. JSME* 25:395 (1982).

10. K.-W. Yi, S. Nakamura, T. Hibiya, and K. Kakimoto, The effect of the Coriolis force on the fluid flow in centrifuge, *in*: "30th National Heat Transfer Symposium of Japan," E342, Yokohama (May 1993), p. 988.

11. K.-W. Yi, S. Nakamura, T. Hibiya, and K. Kakimoto, The numerical study of the Coriolis force on the fluid flow and heat transfer due to wire heating on centrifuge, *Int. J. Heat Mass Transfer* (in press).

CRYSTAL GROWTH OF ENERGETIC MATERIALS DURING
HIGH ACCELERATION USING AN ULTRACENTRIFUGE

M. Y. D. Lanzerotti, J. Autera, J. Pinto

U. S. Army ARDEC
Bldg. 3022
Picatinny Arsenal, NJ 07806 5000

J. Sharma

Naval Surface Warfare Center
White Oak
Silver Spring, MD 20903

ABSTRACT

Studies of the growth of crystals of energetic materials under conditions of high acceleration are reported. This new way of growing crystals of energetic materials by using a concentration gradient is different than the usual procedure of crystal growth by solvent evaporation at constant temperature or by slow cooling. When a solution is accelerated in an ultracentrifuge, the solute molecules concentrate at the outer edge of the tube if the solute is more dense than the solvent. If the solution is initially saturated, then the solution at the outer edge of the tube becomes supersaturated and crystal growth can occur. Results are presented for growth of TNT and RDX crystals at high g in an ultracentrifuge.

INTRODUCTION

Solution crystal growth can be considered a heterogeneous chemical reaction of the type where a portion of the liquid goes into crystal form.[1,2] At 1 g, growth methods include solvent evaporation at constant temperature and slow cooling.[3] Crystal growth occurs when the solution becomes supersaturated; the crystal growth is controlled by simultaneous movement of solute and solvent. Supersaturation can also occur in an initially saturated solution during high acceleration.[4-8] At high g (above 1000 g), crystal growth is controlled by the g-force. The solute molecules individually move through the solvent molecules to form the crystal. A density gradient is established. If the solution is initially saturated, then the solution at the outer edge of the accelerating tube will become supersaturated and crystal growth can occur. Biologists have used a density gradient to separate large biological molecules of slightly different masses.[9] The first international conference on crystal growth at high g addressed many of these issues in numerous papers.[10]

Voids in the accelerated saturated solution migrate out of this saturated solution as a result of the pressure gradient induced by the g-force. Thus bubbles are less likely to form in a crystal grown under acceleration. This feature is important for a number of applications,

Materials Processing in High Gravity, Edited by L.L. Regel
and W.R. Wilcox, Plenum Press, New York, 1994

including those utilizing energetic materials.[11-16] The long term objective of this program is to understand fundamentals of the crystal growth process and thereby to reduce the formation of defects in crystals of energetic materials so they will be less sensitive to mechanical shock.

TECHNIQUE

The experiments on crystal growth are performed using saturated solutions. The samples of saturated solutions are normally filtered prior to insertion into the centrifuge tube in order to remove seed crystals. A Beckman preparative ultracentrifuge model L8-80 with a swinging bucket rotor model SW 60-TI is used to rotate the solution sample up to 60,000 rpm. Both polyallomer centrifuge tubes with hemispherical ends and tubes with hemispherical Teflon inserts to make a flat surface are used in the experiments. After an experimental run the centrifuge tube with saturated solution sample is removed from the bucket and the saturated solution is poured off if a crystal has formed. If necessary, the polyallomer tube is cut lengthwise with a razor blade to study the physical features and habit of the crystal and polycrystal materials without damaging the crystals.

RESULTS

Polycrystalline materials are found on the curved interior surface of the polyallomer centrifuge tube. In the experiments performed to date, the curved surface appears to inhibit single crystal formation. A hemispherical insert with a flat surface interfacing with the saturated solution is inserted into the tube to provide a flat surface that yields single crystal growth.

A number of experimental runs have been made for TNT (trinitrotoluene) and RDX (cyclotrimethylene-trinitramine). These runs have been made for various values of temperature, time, and acceleration. The results are shown in Table 1 and Table 2 for TNT and RDX, respectively.

Table 1. TNT Crystal Growth Experiments At High g and 25°C

Acceleration (g)	Pressure (psi)	Growth Surface	Time (hr)	Filtered Solution	Results
13,000	900	Curved	16	No	No crystals
29,000	2,000	Flat	64	Yes	No crystals
50,000	3,500	Curved	17	No	Polycrystalline
50,000	3,500	Flat	15	Yes	2 individual crystals aligned parallel to acceleration, 2-mm length, habit is coffin-like[17]
50,000	3,500	Flat	92	Yes	Polycrystalline
50,000	3,500	Flat	16	Yes	Individual crystal aligned perpendicular to acceleration, 5-mm length, habit is coffin-like[17]

The results of Table 1 show that 2-5-mm size TNT crystals have been grown from TNT saturated ethyl acetate solution at 50,000 g at approximately 3,500 psi and 25°C for 15 hours. The pressure, p, at the growth surface depends on the density of the TNT saturated solution (≈ 1.4 g/cc) and the acceleration, g. The density of the TNT saturated ethyl acetate solution is estimated from the solubility of TNT in ethyl acetate (59.8 g/100 g ethyl acetate at 21°C),[18] the density of TNT (1.65 g/cc),[19] and the density of ethyl acetate(0.9 g/cc).[20] Acceleration at 50,000 g at 25°C for 92 hours results in polycrystalline TNT. The crystal structure of the 5-mm size TNT crystal has been determined to be monoclinic by x-ray analysis.[17,21]

Table 2. RDX Crystal Growth Experiments At High g

Temp (°C)	Acceleration (g)	Pressure (psi)	Growth Surface	Time (hr)	Filtered Solution	Results
25	50,000	2,000	Curved	17	No	No crystals
0	50,000	2,000	Curved	22	No	Polycrystalline
25	200,000	8,200	Flat	17	No	Individual crystals, 2-mm length, orthorhombic

The results of Table 2 show that 2-mm size RDX crystals have been grown from RDX saturated acetone solution at 200,000 g at approximately 8.200 psi and 25°C for 17 hours. The density of the RDX saturated acetone solution (0.85 g/cc) is estimated from the solubility of RDX in acetone (7.3 g RDX/100 g acetone at 20°C),[18-22] the density of RDX (1.8 g/cc),[19] and the density of acetone (0.79 g/cc).[20]

DISCUSSION

The objectives of this investigation are to understand the fundamental chemistry and physics of crystal growth during high acceleration and to make explosives more insensitive to mechanical shock by reducing the formation of defects in the crystals.[11-16] Good crystals of TNT and RDX have been found to be able to be grown at high g under some conditions of acceleration and temperature. The crystals described herein appear to be free of voids by optical microscopy and are superior to those grown at 1 g by industrial methods or in the laboratory. The individual TNT crystals are coffin-like.[17] The individual RDX crystals are orthorhombic. These initial investigations of the crystal growth of energetic materials at high acceleration have opened a new vista for understanding the formation and structure of these materials.

REFERENCES

1. R. A. Laudise, "The Growth of Single Crystals," Prentice-Hall, Inc., Englewood Cliffs (1970).
2. A. Holden and P. Singer, "Crystals and Crystal Growing," Anchor Books-Doubleday, New York (1960).
3. W. L. Garrett, "The growth of large lead azide crystals", Mat. Res. Bull. 7:949 (1972).
4. P. J. Shlichta and R. E. Knox, Growth of crystals by centrifugation, *J. Crystal Growth*, 3:808 (1968).
5. P. J. Shlichta, Crystal growth and materials processing above 1000 g, *J. Crystal Growth*, 119:1 (1992).
6. I. Amato, The high side of gravity, *Science*, 253:30 (1991).
7. L. L. Regel, Materials Processing In High Gravity, pp. 1-44, Moscow, USSR, 1990.
8. H. Rodot, L. L. Regel, and A. M. Turtchaninov, Crystal growth of IV-VI semiconductors in a centrifuge, *J. Crystal Growth*, 104:280 (1990).
9. O. M. Griffith, "Techniques of Preparative, Zonal, and Continuous Flow Ultracentrifugation," Beckman Instruments Inc. (1986).
10. Proc. First International Workshop on Materials Processing in High Gravity, L. L. Regel, M. Rodot, W. R. Wilcox, eds. *J. Crystal Growth*, 119:1-176 (1992).
11. J. J. Dick, Plane shock initiation of detonation in gamma-irradiated pentaerythritol tetranitrate. *J. Appl. Phys.* 53:6161 (1982).
12. J. J. Dick, Effect of crystal orientation on shock initiation sensitivity of pentaerythritol tetranitrate explosive, *J. App. Phys. Lett.* 44:859 (1984).
13. J. J. Dick, POP plot and Arrhenius parameters for <110> pentaerythritol tetranitrate single crystals, in: "Shock Compression of Condensed Matter-1986," Y. M. Gupta, ed., Plenum Press, New York (1986).
14. J. J. Dick, R. N. Mulford, W. J. Spencer, D. R. Pettit, E. Garcia, and D. C. Shaw, Shock response of pentaerythritol tetranitrate single crystals, *J. Appl. Phys.* 70:3572 (1991).
15. J. Dick, E. Garcia, and D. C. Shaw, Shock initiation of pentaerythritol tetranitrate crystals: steric effects due to plastic flow, in "Shock Compression of Condensed Matter-1991," S. C. Schmidt, R. D. Dick, J. W. Forbes, and D. G. Tasker, eds., Elsevier, Amsterdam (1992).

strength in pentaerthritol tetranitrate, Bull. Am. Phys. Soc., 38:1364 (1993).

17. H. G. Gallagher and J. N. Sherwood, The growth and perfection of single crystals of TNT, in Materials Research Society Symposium Proceedings "Structure and Properties of Energetic Materials", Donald H. Liebenberg, Ronald W. Armstrong, and John J. Gilman, eds., Materials Research Society, Pittsburgh 296:215 (1993).

18. S. Morrow, Growing Explosive Crystals, U. S. Army ARDEC, private communication (1989).

19. B. M. Dobratz and P. C. Crawford, LLNL Explosives Handbook, "Properties of Chemical Explosives and Explosive Simulants," UCRL-52997, Lawrence Livermore National Laboratory, University of California, Livermore, CA, 31 January 1985.

20. R. C. Weast, "Handbook of Chemistry and Physics," CRC Press, Cleveland, 1975-1976 edition.

21. S. M. Kaye, "Encyclopedia of Explosives and Related Items," Picatinny Arsenal Technical Report 2700, 9:T263 (1980).

22. J. T. Rogers, Physical and Chemical Properties of RDX and HMX, Control No. 20-P-26, Holston Defense Corporation, August 1962.

GEL POLYMERIZATION AT HIGH GRAVITY

V. A. Briskman, K. G. Kostarev and T. P. Lyubimova

Institute of Continuous Media Mechanics
Russian Academy of Science
1, Korolyov Street
614061, Perm, Russia

INTRODUCTION

Gravity dependent heat and mass transfer mechanisms often play important roles in material processing. The usage of high gravity conditions for experiments on crystal growth is known. A new field of applications could be the synthesis of polymers. Gel polymerization displays very high gravitational sensitivity,[1-4] and the final sample structure shows traces of all the gel formation stages. There are two main reasons for that; reaction exothermicity and the appearance of a new more dense phase. If the reaction develops nonuniformly, then the above factors lead to density gradients and to macroscopic flows. Transport phenomena directly influence the resulting gel macrostructure. However, the processes forming macro- and microstructures are interconnected. Thus reciprocal influences exist, since the molecular characteristics determine such macroscopic gel parameters as elasticity, permeability, etc.

Due to the above polymerization properties, this process becomes very interesting for a fundamental study of structure formation under high gravity conditions, and very promising from the viewpoint of the development of new technologies. By changing the conditions of the gelatin, it might be possible to control the final sample's properties. In the present work, the influence of high gravity conditions on gel formation was studied theoretically, and experimentally on polyacrylamide gel polymerization in a centrifuge.

MATHEMATICAL MODELING OF GEL FORMATION AT HIGH GRAVITY

Polyacrylamide gel (PAAG) production is usually carried out by crosslinking polymerization of acrylamide in an aqueous solution. The product is a porous matrix. Its structure strongly depends on the total concentration of both comonomers and the percentage of crosslinking monomer. Among other important factors are the solvent viscosity, the temperature, and the reaction initiator concentration. Convection can play an important role in gel polymerization as well.[1-4] The goal of this study was to find out the peculiarities of gel polymerization under high gravity conditions.

Mathematical modeling was developed using the equations of buoyancy convective heat and mass transfer in a photopolymerizing mixture:[5]

$$\frac{\partial \vec{v}}{\partial t} + (\vec{v}\,\nabla)\vec{v} = -\nabla p + \text{Div}\,\hat{\sigma} + \left(Gr_T\theta + Gr_\eta\eta\right)\vec{\gamma} \tag{1}$$

$$\frac{\partial \theta}{\partial t} + \vec{v}\,\nabla\theta = \frac{1}{Pr}\Delta\theta + \frac{FK}{Pr}(1-\eta)^n F(I,H)\exp\frac{\theta}{(1+\beta\theta)} \tag{2}$$

$$\frac{\partial \eta}{\partial t} + \vec{v}\,\nabla\eta = \frac{1}{Sc}\Delta\eta + \delta\frac{FK}{Pr}(1-\eta)^n F(I,H)\exp\frac{\theta}{(1+\beta\theta)} \tag{3}$$

$$\text{div}\,\vec{v} = 0 \quad, \quad \hat{\sigma} = H\hat{e} \quad. \tag{4}$$

Here $\vec{v}$ is dimensionless velocity, p the convective part of the dimensionless pressure, θ the dimensionless temperature, η is the conversion, $\hat{\sigma}$ and $\hat{e}$ are the viscous stress and shear rate tensors, $F(I,H)$ describes the dependence of the effective reaction rate on the intensity of irradiation, I, and the effective viscosity H, and n stands for the order of the reaction.

Dimensionless parameters are Grashof numbers, Gr_T and Gr_η, Prandtl number Pr, Schmidt number Sc, parameters δ and β, and the Frank–Kamenetzky parameter FK:

$$Gr_T = \frac{\rho^2 g\,\beta_T RT_0^2 L^3}{\mu^2 E_{ef}} \quad;\quad Gr_\eta = \frac{\rho^2 g\,\beta_\eta L^3}{\mu^2} \quad;\quad Pr = \frac{\mu}{\rho x} \quad;\quad Sc = \frac{\mu}{\rho D_f} \quad;$$

$$\delta = \frac{\rho\,C_p RT_0^2}{Q\,E_{ef}} \quad;\quad \beta = \frac{RT_0}{E_{ef}} \quad;\quad FK = \frac{Q\,E_{ef}L^2 K_{ef}^0 \exp\left(-E_{ef}/RT_0^2\right)}{\kappa\,RT_0^2} \quad. \tag{5}$$

We assumed that the effective viscosity of the reacting mixture does not remain constant, but changes during polymerization:

$$H = (1 + A\eta)^m \exp(-\gamma\theta) \quad. \tag{6}$$

The effective rate of photopolymerization was assumed to depend on the intensity of irradiation and the viscosity according to the law:

$$F(I,H) + \begin{cases} I_d^{1/2} & , \text{ if } H < H_c \\ I_d^{1/2}(H/H_c)^{1/2} & , \text{ if } H \geq H_c \end{cases} \tag{7}$$

where I_d is the local intensity of irradiation at distance d from the irradiated surface, as defined by:

$$I_d = \exp(-Bd) \quad . \tag{8}$$

Here, B is a dimensionless parameter characterizing the absorption of irradiation by the solution.

The calculations were made for a rectangular parallelepiped with one of the sides being significantly shorter than the others (a Hele–Shaw cell). In this case, it is possible to use the approximation of plane trajectories and to solve the problem using a two–dimensional formulation.

On the rigid sidewalls of the cell, the no–slip conditions, the linear law of heat transfer, and the impermeability condition have been imposed, while external heating was absent:

$$\vec{v}\big|_\Gamma = 0 \quad , \quad \frac{\partial \theta}{\partial n}\bigg|_\Gamma = -\alpha\theta \quad , \quad \frac{\partial \eta}{\partial n}\bigg|_\Gamma = 0 \tag{9}$$

Polymerization kinetics can be described with a phenomenological approach.[1-8] It is simple and contains few parameters. A more complete description of the process can be obtained with a kinetic approach. We used the stationary state hypothesis for radicals and the long chain approximation, and neglected physical crosslinking and termination by disproportionation. We derived the following kinetic equations:[5]

$$\frac{d[A]}{dt} = -k_i[R*][A] - k_p\big([\dot{P}] + [\dot{V}]\big)[A] \tag{10}$$

$$\frac{d[B]}{dt} = -2k_i[R*][B] - 2k_p\big([\dot{P}] + [\dot{V}]\big)[B] \tag{11}$$

$$\frac{d[U]}{dt} = 2k_i[R*][B] + 2k_p\big([\dot{P}] + [\dot{V}]\big)[B] - k_i[R*][U] - k_p\big([\dot{P}] + [\dot{V}]\big)[U] \tag{12}$$

$$\frac{d[S]}{dt} = k_p\big([\dot{P}] + [\dot{V}]\big)[U] \tag{13}$$

$$[\dot{P}] + [\dot{V}] = \left(\frac{k_i[R*]}{k_t}\big([A] + 2[B] + [U]\big) \right)^{1/2} \tag{14}$$

Here, $[A]$ and $[B]$ stand for the concentrations of comonomers, $[S]$ is the crosslinker's concentration, $[\dot{P}]$ and $[\dot{V}]$ are the concentrations of polymer radicals, and k_i, k_p and k_t are the rate constants for initiation, propagation and termination. The termination rate constant k_t is a function of the effective viscosity defined according to Eq. (6). $[R*]$ is a function of the local intensity of the irradiation defined in accordance with Eqs. (7) and (8). The temperature dependencies of the rate constants are given by the Arrhenius law.

A compatible solution of Eqs. (10) to (14) and the equations for motion and heat transfer were carried out using the kinetic approach. A finite–difference method was used processes, the combination of the Lagrange and Euler approaches, and the predictor-corrector method with a variable time step. The calculations were made for the direction of irradiation through the narrow sidewall.

As one can see from Eqs. (7) and (8), the rate of photopolymerization is proportional to the square root of the light intensity absorbed by photoinitiator. This value is different at different distances from the irradiated surface. In microgravity conditions, when convective heat and mass transfer are negligible and the process is controlled by chemical kinetics, this leads to the distribution of the conversion and the crosslink density being characterized by distinct gradients along the light beam line all over the gel matrix. The highest values of the conversion and the crosslink density are reached near the irradiated surface (see Fig. 1a, where the corresponding distribution of the crosslink density is presented). Since the elastic modulus in gels is proportional to the effective crosslink density,[6,7] the distribution of the elasticity would look like the field described in Fig. 1a. Due to the correspondence between elastic modulus and pore size,[6,7] one could expect a gradient distribution of the pore size as well.

Figure 1. Effective crosslink density isolines for photopolymerization under (a) microgravity, (b) 1 g, and (c) high gravity conditions.

Sedimentation and convective heat and mass transfer occur when polymerization is carried out at 1 g. Since the new (polymer) phase is more dense than the initial one, the new phase particles formed near an irradiated surface move down in the gravitational field. Near the cavity bottom, this results in conversion and crosslink density higher than in the other sections. There the properties are graded along the gravity direction. Mixing by convection in the remaining part of the liquid, where the conversion and the effective viscosity are lower, leads to gel nonuniformity there (see Fig. 1b).

Under high gravity the gravitational effects are very large. Most of the new phase particles formed near an irradiated surface move to the bottom. This results in gradients of the conversion and the crosslink density in most of the gel matrix (see Fig. 1c). This demonstrates the possibility of synthesizing a gel matrix of graded properties at high gravity.

EXPERIMENTAL STUDY OF THE INFLUENCE OF CENTRIFUGAL FORCE ON GELATION

The influence of centrifugal force on gel structure was studied experimentally. Photoinitiated polymerization was carried out in a 15% aqueous solution of acrylamide in the presence of a crosslinker. Riboflavin was used as the photoinitiator.

There are two important factors in the photopolymerization process that are influenced by centrifugal force. The first one is the difference in the reaction rate at different distances from an irradiated surface. Since the polymer phase is more dense than the monomer, this creates conditions for convection. The second effect is sedimentation. The more dense polymer particles move to the periphery. These two effects can markedly change the final gel structure.

The experimental model was a horizontal plane disk of 5.5 mm thickness and 60 mm in diameter filled by the reaction mixture (see Fig. 2). Acrylic plastic transparent plates formed the top and bottom surfaces (1). The model was fixed on a heavy steel face plate (2) mounted on a motor shaft. The axis of rotation of the model coincided with that of the shaft. The deviation of the axis of rotation from the vertical was less than 30 arc sec. The rotational speed of the motor was 3000 revolutions per minute.

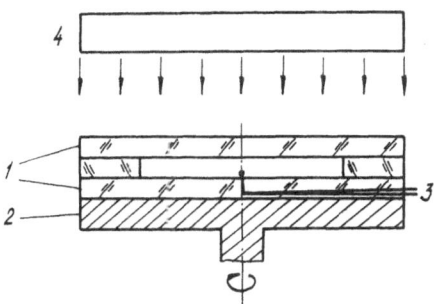

Figure 2. Experimental model: 1 – transparent plates, 2 – steel face plate, 3 – differential thermocouple, and 4 – photoinitiating light source.

Tests were performed at room temperature (293–295 K). The reaction of PAAG polymerization is exothermal, so to measure heating of the reaction mixture at the center of the cavity, a copper–constantan thermocouple (3) was used. Free convection of air and heat transfer to the face plate through the bottom surface allowed removal of heat from the model. A special light source with maximal radiation intensity at 445 nm (4) was used for photoinitiation. The light source was placed above the model at a height of 15 cm.

The structure of gel matrices was studied as in the previous work;[4] that is, cylindrical samples of radius 4.5 mm and height 5.5 mm were cut from a gel matrix polymerized for one hour. Samples were taken at progressive distances from the disk center, so that it would be possible to study the elastic modulus distribution along the radius. Due to the relationship between elastic modulus and pore size, these measurements give information on the pore size as well.

The change in gel structure under the influence of external factors, including acceleration, depends markedly on the polymerization stage at which the initiation of the process starts. There is a relationship between the gelation stages and the shape of the temperature curve characterizing the heating of the material. Figure 3 shows the shape of the temperature curve for the experimental model. The section of the curve AB corresponds to reaction initiation; section BC demonstrates gel formation from the liquid phase. Under photoinitiation, further polymerization takes place in the polymer gel phase, indicated by section CD.

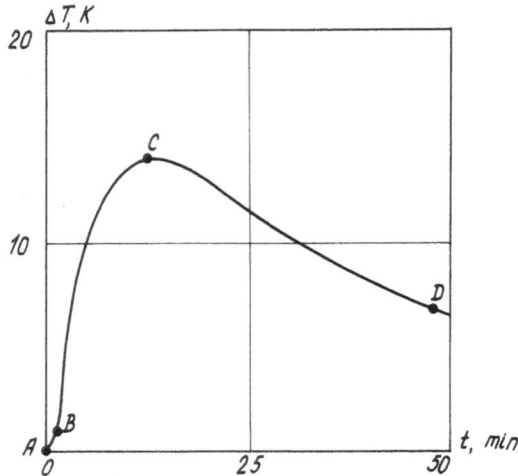

Figure 3. Temperature increase in reaction mixture during polymerization.

The temperature curve in Fig. 3 was obtained during polymerization at rest. In this case, the elastic properties in the gel sample did not change along the radius; the elastic modulus equals 4.05×10^4 N/m^2.

The situation was different when the model was rotated. In that case, the elastic properties varied along the disk radius. In Fig. 4 Young's modulus versus radial coordinate is presented for different time intervals t_0 between switching on rotation and the start of initiation. Curves 1 and 2 correspond to t_0 equal to 0 and 8 minutes. As one can see, the dependence of the elastic modulus on radius is linear. These gel samples, as well as the other ones formed in the interval BC (see Fig. 3), are ring shaped. The reason is that in these cases rotation was started while the reaction mixture was still liquid. High acceleration led to the separation of the mixture – more dense polymer particles moved to the periphery of the rotating disk. The central part of the cavity was occupied by water.

Gel samples polymerized in the interval CD were of different structure. In these cases the rotation was started when the processes occur in the gel phase. Centrifugal forces change the gel sample so that the less dense phase (solvent) saturates the pores in the central part of the disk, while the more dense polymer chains move to the periphery. This can be the reason for the linear parts of curves 3 and 4 in Fig. 4 (t_0 values 12 and 18 min).

The further increase of acceleration with increasing distance from the rotation axis leads to the partial mechanical destruction of the gel sample. This causes a decrease of elastic

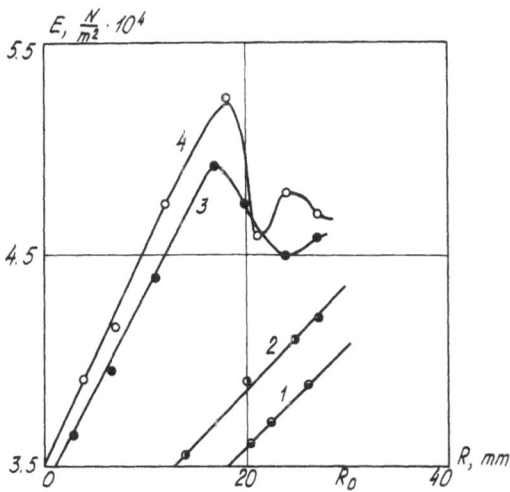

Figure 4. Elastic modulus distribution over the disk radius for matrices polymerized in the presence of centrifugal force, t – period between the beginning of photoinitiation and the start of rotation: 1–0, 2–6, 3–12, and 4–18 min.

modulus in that interval (right parts of curves 3 and 4 in Fig. 4). Due to the presence of defects, the correspondence between elastic modulus and pore size no longer exists; so that is impossible to make any conclusions about the pore size distribution in those parts of the sample.

CONCLUSIONS

Mathematical modeling was developed to describe gel formation at high gravity. It was shown that these conditions strongly influence gel formation due to reaction nonuniformity and the appearance of the more dense polymer phase. The possibility to synthesize gel matrices with graded properties by photopolymerization under high gravity was demonstrated.

Experiments were carried out to study gel formation in the presence of centrifugal force. Elastic measurements were performed for analysis of the final gel structure. The gels synthesized in the rotating model had different properties than those obtained in the rest of the chamber. The effect strongly depended on the reaction stage when the rotation was switched on. In all cases there was an area with graded properties in the gel matrix.

Thus, it was demonstrated both by numerical investigation and by experiments in a rotating cell, that high gravity provides the possibility to synthesize gels with new, graded structure.

The same physical phenomena might exist and play an important role in processing other multiphase and multicomponent media. The very high gravitational sensitivity of gels make them useful for modeling the influence of high gravity on structure formation.

REFERENCES

1. A.S. Sadykov, V.B. Leontyev, Yu.S. Mangutova, G.M. Grechko, G.S. Nechitailo, and A.L. Mashinskii, *Akad. Nauk SSSR, Doklady* 303:1004 (1988).

2. Sh.D. Abdurakhmanov, L.G. Bogateyreva, V.A. Briskman, M.G. Levkovich, V.B. Leontyev, T.P. Lyubimova, A.L. Mashinskii, and G.S. Nechitailo, On polyacrylamide gel formation by photoinitiation under terrestrial and orbital conditions, *in*: "Numerical and Experimental Modelling of Hydrodynamic Phenomena under Weightlessness," Sverdlovsk (1988).

3. Sh.D. Abdurakhmanov, V.G. Babskii, L.G. Bogatyreva, V.A. Briskman, M.G. Levkovich, V.B. Leontyev, T.P. Lyubimova, A.L. Mashinskii, and G.S. Nechitailo, Structure formation of polyacrylamide gel at photoinitiation under earth and orbital conditions, *in:* "Gagarin Scientific Readings on the Astronautics and Aeronautics 1989," Moscow (1990).

4. L.G. Bogatyreva, V.A. Briskman, K.G. Kostarev, V.B. Leontyev, M.G. Levkovich, T.P. Lyubimova, A.L. Mashinskii, G.S. Nechitailo, and P.G. Righetti, Heat/mass transfer mechanisms of the polymerization under terrestrial and microgravity conditions, *in:* "Proceedings of the VIII European Symposium on Material and Fluid Sciences in Microgravity," ESA SP–333, Vol. 1 (1992).

5. T.P. Lyubimova, Polymerization under terrestrial and orbital conditions: comparative study, *in:* "Hydromechanics and Heat/Mass Transfer in Microgravity, Reviewed Proceedings of the First International Symposium on Hydromechanics and Heat/Mass Transfer in Microgravity, Perm–Moscow, 1991," Gordon and Breach Science Publishers (1992).

6. P. deGennes. "Scaling Concepts in Polymer Physics" (1979).

7. T. Lyubimova, S. Caglio, C. Gelfi, P.G. Righetti, and Th. Rabilloud, Photopolymerization of polyacrylamide gels with methylene blue, *Electrophoresis* 14:40 (1993).

NUMERICAL SIMULATION OF THE EFFECT OF GRAVITY ON WELD POOL SHAPE

J. Domey,[1] D.K. Aidun,[1] G. Ahmadi,[1] L.L. Regel,[2] and W.R. Wilcox[2]

[1]Mechanical and Aeronautical Engineering Department
[2]International Center for Gravity Materials Science and Applications
Clarkson University
Potsdam, NY 13699

ABSTRACT

Understanding the physical phenomena involved in the welding process is of substantial value to improving the weldability of materials. The intense heat and the arc inherent in fusion welding make direct experimental observation of the weld pool behavior rather difficult. Thus numerical models that can predict the processes involved have become an invaluable tool for studying welding.

One of the major factors affecting the motion within the molten weld pool is the gravity-driven buoyancy force. This force can act to oppose or enhance the Marangoni convective flow within the weld pool. To study the effect of gravity on weld pool processes, a series of numerical simulations was performed. It was found that higher gravitational fields tend to enhance the convective flow within the weld pool and thus affect the heat transfer, the depth and width of the two phase region, and the pool depth-to-width ratio.

INTRODUCTION

According to David and Debroy,[1] "Losses of life and property damage due to catastrophic failure of structures are often traced to defective welds." Since welding is such an important and widespread fabrication technique, it is imperative that a basic understanding of the physical processes involved become available. Until recently, welding has been thought of more as an art than a science with techniques developed through trial and error methods. Since welding is used in such a wide range of metal joining applications, from bicycles to nuclear reactor cores, providing a fundamental understanding of processes involved is of crucial interest to many fields. David and Debroy[1] state that "Reliable science based correlations between the microstructure and properties of welds as well as models to predict such relations are important for the development of the field."

Materials Processing in High Gravity, Edited by L.L. Regel
and W.R. Wilcox, Plenum Press, New York, 1994

Several problems arise that cause defects within welds. One major source of defects is hot cracking. Factors affecting the hot-cracking susceptibility of an alloy fall into two categories; metallurgical and mechanical. The metallurgical factors are controlled by the composition and solidification morphology of the weldment. The mechanical factors are controlled by thermal stresses and strains. These mechanical factors occur in the material as it goes through its intense thermal cycle that causes the metal to solidify rapidly. Since these cracks are detrimental to the quality of the weldment, and ultimately the work piece, it is beneficial to be able to develop a proper welding procedure so that hot cracking can be avoided. In order to achieve this goal, it is first necessary to understand the physical processes that occur during welding.

The nature of arc welding does not allow direct observation during the welding process, and physical observation of the weld is limited to solidified welds. Thus, computational simulations are needed to provide a better understanding of the transient phenomena that are present during the welding process.

BACKGROUND

A brief review of the published literature on weld pool modelling is provided in this section. Most earlier models developed for the prediction of weld pool characteristics are limited by many simplifying assumptions. Some of the common restrictive assumptions that cause the models to be unrealistic include a prescribed weld pool profile, an undeformable weld pool surface, a stationary heat source, and a two-dimensional (2D) simplification of a three-dimensional (3D) problem. Only the model developed by Zacharia et al.[2] relaxes many of these limitations and provides a realistic computational model for calculating weld pool characteristics.

In 1988 Zacharia, Eraslan, and Aidun[2,3] developed a 3D transient model to simulate a moving gas-tungsten-arc (GTA) and a gas-metal-arc (GMA) welding process. The model incorporates a deformable surface and allows for mass addition and surface evaporation. The researchers found that the Marangoni forces were dominant, that surface deformation could retard Marangoni effects, and that the predicted surface deformation was in agreement with observation. Since the surface is completely deformable, the model accurately describes the phenomena of surface rippling and the weld "crown." In addition the code may be applied to welding in microgravity conditions.

Tsao and Wu[4] presented a transient model that simulates both GMA and GTA welding processes. With their model they were able to account for the thermal energy addition of the filler metal within the arc. By applying this model to both GMA and GTA welding, they found that GMA penetrates three times faster than GTA under similar conditions. They also found that the surface properties have little influence on the GMA weld pool shape. The main influence on the increased weld penetration results from the molten filler metal (spray).

A 2D transient model of the GTA weld process was formulated by Thompson and Szekely.[5] This model does not consider the effect of Lorenz forces, but does allow for the depression of the free surface. The model showed that the depression of the free surface can affect the maximum surface velocity by up to about 10% and can cause a change in the weld penetration about equal to the amount of the depression.

Tsotridis, Rother, and Hondros[6] constructed a 2D transient model that simulates a laser weld process and was the first to account for heat losses due to surface evaporation. They found that Marangoni forces are dominant and that the weld profile near the surface is partially dependent upon the power of the laser beam.

Pardo and Weckman[7] developed a 3D finite-element model for the calculation of steady-state temperatures in the GMA welding process. The model accounts for the release of the latent heat of fusion at the solid-liquid interface and is capable of predicting the weld reinforcement geometry. This model can also handle non-uniform velocity fields and surface geometries within the liquid weld pool.

Recently, Zacharia and co-workers[8,9] used metallographic techniques to compare the actual fusion zone geometry of laser and GTA welds onto 304 stainless steel with their numerical simulation results. The model correctly predicts the shape of the fusion zones. They found that the weld bead is obtained as a result of the solidification of the liquid metal so that the behavior of the liquid metal during solidification in the fusion zone should be considered an essential influence on the final properties of the weld.

In the present work, numerical simulations of GTA welds onto the aluminum alloy 6061, are presented for varying levels of gravity. The WELDER code is used for the theoretical prediction of the heat transfer processes involved in a standard GTA weld. The model considers the buoyancy, electromagnetic, and surface tension forces when solving for the overall heat transfer for a work piece of finite size and shape. The model also accounts for phase change and considers the temperature dependence of the thermophysical properties. The relevant thermophysical properties for 6061 aluminum and the appropriate GTA welding process conditions are utilized in the simulation so that accurate results are obtained. The effects of gravity on the convection patterns and thermal conditions in stationary and moving weld pools are studied. The consequences of gravity on weld pool depth-to-width ratio is also discussed.

NUMERICAL SIMULATION

Overview of WELDER Code

The WELDER code is a transient, three-dimensional computer simulation model which was developed for the investigation of coupled conduction and forced- and natural- convection heat transfer problems associated with the welding process. On the basis of modeling of physical phenomena, the special features of the code include: (1) realistic treatment of the molten surface of the weld pool as a deformable surface, (2) detailed consideration (without resorting to the Boussinesq approximation) of all of the densimetric-effect terms, (3) detailed consideration of the electromagnetic force effects, (4) accurate treatment of the mass addition to the weld pool (non-autogenous welding), (5) accurate treatment of the transient shape of the solid-liquid interface, according to a non-equilibrium (kinetic) phase-change model, (6) correct treatment of the combined surface-tension pressure and surface-tension-gradient effect (Marangoni shear-stress effect), (7) consideration of an arbitrary gravitational force (both low and high g), (8) consideration of the inclination of the workpiece relative to the gravitational force field (simulating out of position welding), (9) detailed consideration of surface cooling (convection and radiation), (10) realistic treatment of surface evaporation of the metal in the weld pool, and (11) accurate representation of the moving arc conditions (linear welds).

Special computational features include: (1) geometrically accurate composite-space-splitting discretization algorithm of the discrete-element-analysis, (2) composite-time-splitting explicit integration algorithm, with directional-transportive-upwind interpolation, which guarantees the stability of numerical solutions with second-order accuracy, and (3) marked-element formulation for accurate computation of the transient solid-liquid interface of the two-phase mushy-zone subregion.

For a more complete description of the WELDER code, along with the discretization algorithm, one is referred to the work of Eraslan et al.[3] and Domey.[10]

Buoyancy Driven Flow

The densimetric-coupling associated with the variation of the density of the liquid metal is included in the WELDER code. The local density of the liquid metal is considered as a constant reference value plus a generalized compressibility factor which represents the percent density variation with temperature.[2] That is:

$$\rho = \rho_o (1 + \frac{\Delta\rho}{\rho_o}) = \rho_o (1 + \beta) \tag{1.a}$$

$$\beta = \beta(T) = \frac{\Delta\rho}{\rho_o}(T) \tag{1.b}$$

where ρ is the local density, ρ_o is the reference density, β is the compressibility factor, and T is the temperature.

The gravitational force has a direct effect on the flow within a weld pool (through the buoyancy effect) and can be used to either enhance or deter the flow of molten material. When a fluid goes through a temperature change, there is also a corresponding change in its density. For welding, the incident heating upon the surface of the molten weld pool causes the melt to rise in temperature and, thus, go through a change in density. For most cases, the density decreases as the temperature increases.

A schematic of a buoyancy driven flow pattern is shown in Figure 1. This figure shows how the temperature gradient within a weld pool causes a corresponding density gradient and enhance the flow. When material of a higher temperature and lower density is forced to the bottom of a weld pool, the buoyancy force causes it rise up through the center of the pool. The flow moves radially outward, the hot material is forced along the surface and then down the sides of the weld pool to the bottom. This leads to the circulation flow pattern shown in Figure 1. The buoyancy-driven convection tends to decrease the depth-to-width ratio.

Earlier the WELDER code was used by Domey[10] and Domey et al.[11] to study welding of 6061 aluminum and the titanium alloy, Ti-6Al-4V. Their results showed that the WELDER code is a suitable tool for the investigation of heat transfer phenomena involved in GTA welding.

Numerical Parameters

The simulations were performed for a stationary 150 amp, direct-current-electrode-negative (DCEN), 21 volt, GTA weld into a 24 x 24 x 6mm workpiece of the aluminum alloy 6061. This alloy was chosen due to its widespread use in the aerospace industry. In order for

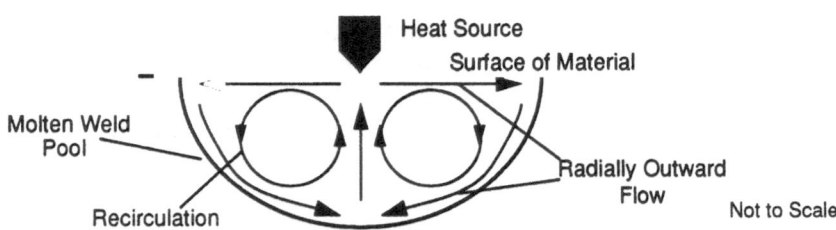

Figure 1. Schematic of Buoyancy Driven Flow

Table 1. Thermophysical Properties and Weld Parameters[12]

Property or Parameter	Value
Solidus Temperature	582 °C
Liquidus Temperature	652 °C
Vaporization Temperature	2467 °C
Reference Density	2700 kg/m^3
Solidus Density	2292 kg/m^3
Liquidus Density	2241 kg/m^3
Magnetic Permeability	1.26 (10^{-6})H/m
Latent Heat of Fusion	1516 kJ/kg
Latent Heat of Vap.	40138 kJ/kg
Liquidus Viscosity	0.375 (10^{-6})kg/m•s
Thermal Diffusivity	37.5 (10^{-6})m^2/s
Specific Heat, C_p	1.066 kJ/kg/K
Surface Tension @ Liquidus	9.14 (10^{-6})N
Surface Entropy	-3.5 (10^{-11})N/°C
Arc Current	150 Amps DCEN
Arc Efficiency	100 %
Arc Power	3.153 kJ/s
Room Temperature	25.0 °C
Effective Radius of Heat Flux	0.003 m

a simulation to predict accurate results, all of the relevant thermophysical properties for the given material must be known. The values for the thermophysical properties used in the present simulations are listed in Table 1.

Boundary Conditions

As noted before, the specimen was assumed to be a 24 x 24 x 6 mm piece of 6061 aluminum, as shown in Figure 2, and was assumed to be completely surrounded by air at room temperature. This figure is a schematic diagram of the system that was used in both of the 6061 aluminum simulations. The electrode was placed directly over the center. Natural convection with the surroundings was assumed at the boundaries, with evaporation allowed from the liquid surfaces. For both of the stationary GTA weld simulations, the electrode was placed directly above the center of the surface of the specimen and held stationary throughout the simulation. The gravitational force was assumed to be acting normal to the surface of the workpiece and directed downward, (the negative 'z' direction).

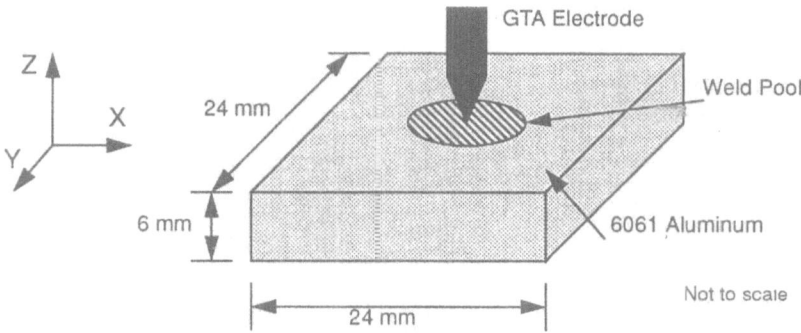

Figure 2. Schematic diagram of the workpiece and the weld pool

Grid System

The grid system employed in both the stationary and linear gas-tungsten-arc weld (GTAW) simulations was a 16 x 16 x 8 grid, as shown in Figure 3. This figure shows the numerical grid system that was used in the 6061 aluminum simulations as the geometry was broken down into 16 divisions in each of the X and Y directions and 8 divisions in the Z direction. The divisions are smaller in the center and near the top where most of the "action" takes place. The spacings in the central area, where the weld pool forms, are smaller to improve resolution. Although this grid was relatively coarse, it provided results for quantitative analysis, while keeping the computational time within an acceptable limit. Should a finer mesh for a higher accuracy be desired, it could easily be implemented at the expense of additional computational time.

A SPARC-station 2 GX, along with a 600 MB hard drive, was used to provide the necessary computational power and storage capability. The resulting data files from WELDER were plotted using Tecplot 5.0 and then printed using a PostScript laser printer.

RESULTS

Several simulations were performed for different levels of gravity. The results of three of the simulations are shown in Figure 4, which shows both the top views as well as the side views for the 0.1g, 1.0g, and the 2.0g simulations. As can be seen from the figure, rather complex convection patterns are formed. The higher g produces greater velocities, as indicated by longer arrows.

For the 1g and 2g cases, the flow was radially outward. For the 0.1g case the flow was radially inward. This showed that for smaller g, the surface tension driven (Marangoni) force dominates, whereas buoyancy forces dominate at normal or high gravity conditions. In high g, the thermally driven buoyancy force near the center of the pool was sufficiently large to overcome the surface tension force and the convection pattern reverses. This was particularly evident for the high gravity case of 2g in Figure 4.

Since the buoyancy driven flow was radially outward at the surface, one would expect that the weld pool would expand in width at a greater rate than it would in depth, thus creating a larger depth-to-width ratio.

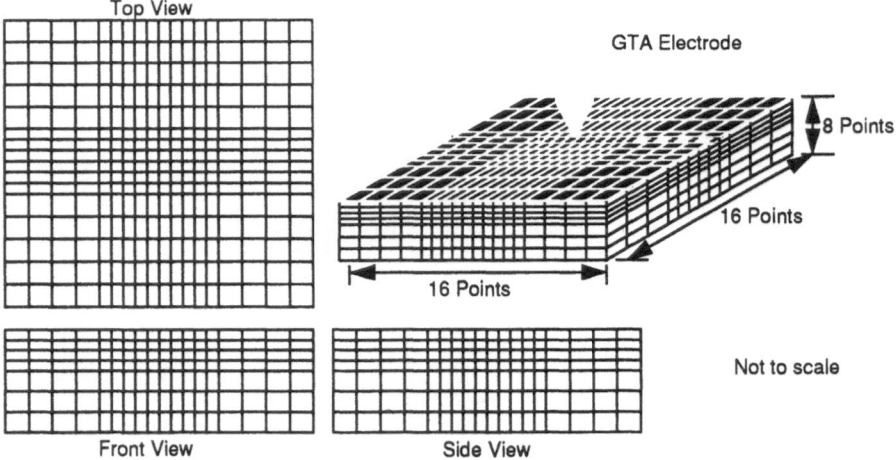

Figure 3. Grid system used in simulation

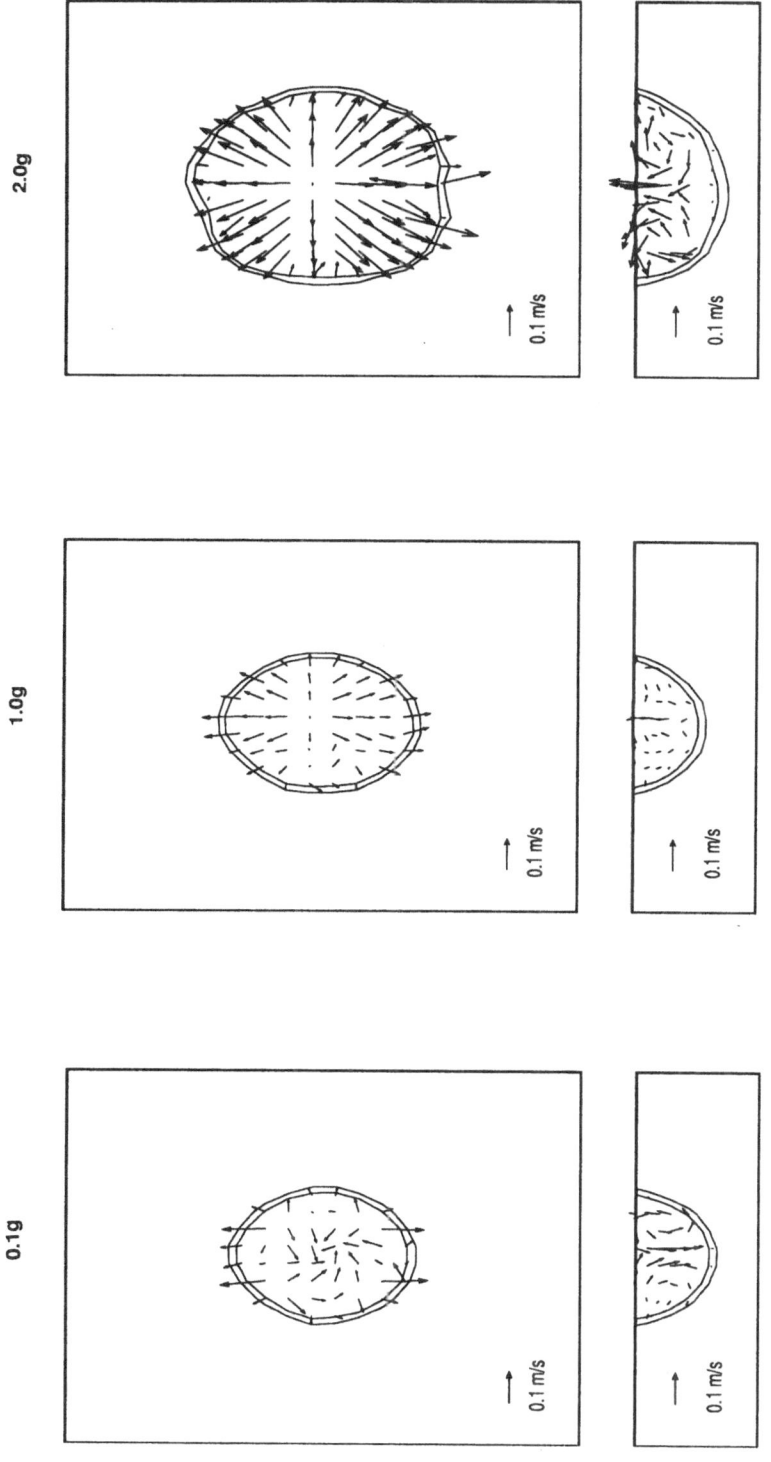

Figure 4. Stationary GTA welds at different gravity levels - (g=earth's gravity)

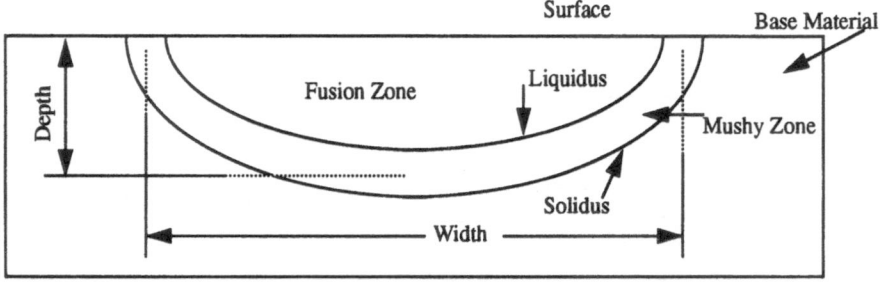

Figure 5. Definitions of depth, width, and mushy zone

Depth, Width, and Depth-to-Width Ratio

In this work, the depth of the weld pool was defined as the distance normal to the surface of the weld pool down to a point midway between the solidus and liquidus isotherms (see Figure 5). Similarly, the width was defined as the distance along the surface of the weld pool, as measured from the midway points between the solidus and liquidus isotherms across the diameter of the molten pool.

The depth, width, and depth-to-width ratio (d/w) are plotted versus gravity in Figure 6. As can be seen from the figure, high gravity causes both the depth and width of the weld to increase. The slope of the width line was greater than that of the depth line. This can also be seen by noting that the depth-to-width ratio has a slight negative slope, indicating that the gravitational effect on the width was greater than on the depth.

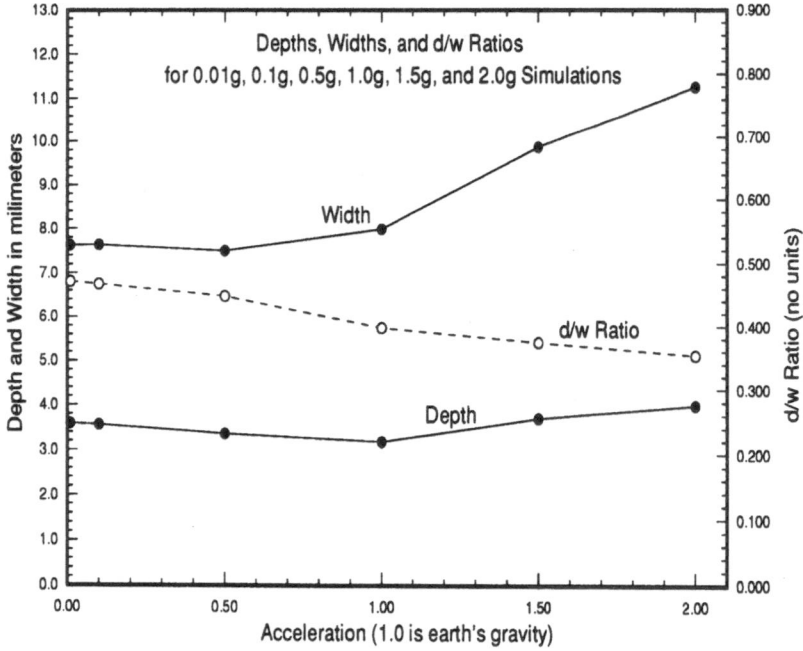

Figure 6. Depth, width, and depth-to-width ratio versus gravity

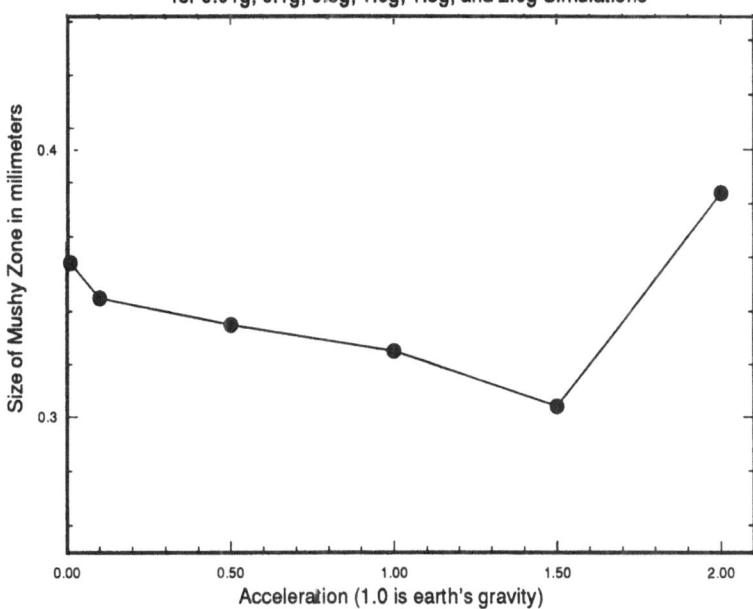

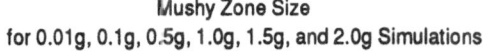

Figure 7. Mushy zone size versus gravity

Mushy Zone

The mushy zone, or two phase region, was defined as the region bounded by the solidus and liquidus lines, as shown in Figure 5. Although these isotherms are derived from the equilibrium phase diagram and a weld is a transient process, this definition allows for a quantitative comparison between simulations to be drawn. In Figure 7 it can be seen that as the gravitational force was increased, the mushy zone size was slightly decreased until an acceleration of 1.5g was reached. At this point the mushy zone size begins to increase indicating that the solidus isotherm was expanding into the base material faster than the liquidus isotherm. This phenomena is currently under investigation.

SUMMARY

Numerical simulations of a GTA welding process in aluminum alloy 6061 were performed for different levels of gravity. The results show that a high gravitational field causes an enhanced buoyancy-driven radially outward flow in the weld pool. This rather high speed flow causes an increase in the heat transfer as compared to the lower gravity cases. The increased heat transfer affects the depth, width, d/w ratio, and the size of the mushy zone. The depth and width both increase with an increase in the gravitational field, but the width grows more rapidly, resulting in a decreased d/w ratio. The mushy zone also increases slightly with a corresponding increase in the gravitational field.

REFERENCES

1. S.A. David and T. Debroy, "Current Issues and Problems in Welding Science," *Science* 257:497 (1992).
2. T. Zacharia, A.H. Eraslan, and D.K. Aidun, *Welding J.* 67:18s (1988).

3. A.H. Eraslan, T. Zacharia, and D.K. Aidun, Report No. MIE-142, Clarkson University, Potsdam, NY (1986).

4. K.C. Tsao and C.S. Wu, *Welding J.* 67:70s (1988).

5. M.E. Thompson and J. Szekely, *Intl. J. of Heat and Mass Transfer* 32:1007 (1989).

6. G. Tsotridis, H. Rother, and E.D. Hondros, *Naturwissensch* 76:216 (1989).

7. E. Pardo and D.C. Weckman, *Met. Trans. B* 20B:937 (1989).

8. T. Zacharia, S.A. David, J.M. Vitek, and T. Debroy, *Welding J.* 68:499s (1989).

9. T. Zacharia, S.A. David, J.M. Vitek, and T. Debroy, *Welding J.*, 68:510s, (1989).

10. J. Domey, "Digital Simulations of a Stationary and a Linear Weld," Report No. MAE 269, Clarkson University, Potsdam, NY (1993).

11. J. Domey, D. Aidun, G. Ahmadi, T. Diebold, "Numerical Simulation of GTA Welds on Titanium Alloys with Comparison to Experimental Results," Proceedings to the 3rd International Conference on Trends in Welding Research, Gatlinburg, TN (1992).

12. T. Zacharia and D.K. Aidun, "Fundamental Investigations of Al-Li-Cu-Mg Alloy Weldments," Report No. MIE-151, Clarkson University, Potsdam, NY (1987).

HIRB - THE CENTRIFUGE FACILITY AT CLARKSON

Ramnath Derebail,[1] William A. Arnold,[2] Gary J. Rosen,[1]
William R. Wilcox,[1] and Liya L. Regel[1]

[1]International Center for Gravity Materials Science and Applications
Clarkson University
Potsdam, NY 13699
[2]NASA Lewis Research Center
Lewis Research Center, MS105-1
21000 Brookpark Road
Cleveland, OH 44135

ABSTRACT

The International Center for Gravity Materials Science and Applications at Clarkson University has constructed the only centrifuge facility in the world dedicated solely to materials processing research and related flow visualization. This centrifuge has been named "HIRB," which is an acronym for "High Inertia Rotating Behemoth." HIRB is a modified variable-speed boring mill with an arm radius of 1.524 m. The maximum attainable rotation rate is 90 rpm, which corresponds to a maximum acceleration of approximately 13.8 g (g = 9.81 m/s^2). This paper will focus on the main characteristics of HIRB and its capabilities for gravity materials science experiments. Equipment used for gravity materials processing experiments on HIRB are described. A low cost, low weight and low power gradient freeze furnace was designed, constructed and tested on HIRB under high gravity conditions.

INTRODUCTION

Large centrifuges are mainly used in civil engineering applications. The primary function of the 18 m arm centrifuge used by the Soviets to perform crystal growth experiments at Star City near Moscow was to train cosmonauts to withstand high acceleration during take off and reentry of spacecraft[1,4,5] . The 5.5 m arm centrifuge used by the French to perform crystal growth experiments at Nantes in France belongs to the Department of Roads and Bridges and is used mainly for civil engineering studies

Materials Processing in High Gravity, Edited by L.L. Regel
and W.R. Wilcox, Plenum Press, New York, 1994

on small scale models[2,3]. Around the world, considerable difficulties are encountered by material scientists to obtain time on centrifuges to perform crystal growth experiments. Even when access to centrifuges was obtained, the time available was very limited and modifications had to be done to perform materials processing experiments. Therefore, to overcome these problems, we at the International Center for Gravity Materials Science and Applications decided to construct a centrifuge facility dedicated solely to materials processing research.

BACKGROUND

Large centrifuges are available commercially. However the main drawback is that these centrifuges are designed for civil engineering applications. Furthermore they are expensive, with prices ranging from $150,000 to $10,000,000 or more. Due to these high costs, we decided to construct our own centrifuge by modifying an existing machine. This machine was a boring mill. The primary function of a boring mill is to machine large circular parts.

The boring mill that we purchased was built in 1949 by King Machine Tools of Cincinnati, Ohio. This boring mill was installed at a naval shipyard and used to machine bomb shells, torpedo casings, etc. The Akron Equipment Company of Akron, Ohio salvaged this boring mill from the naval shipyard about 30 years ago. It was primarily used to machine tire molds for Goodyear Tire Company. Clarkson University purchased this boring mill from Akron Equipment Company. It was then modified to run as a centrifuge and christened "HIRB," which is an acronym for "High Inertia Rotating Behemoth." Figure 1 is a photograph of the boring mill before any modifications were made.

Figure 1. Photograph shows the boring mill before modifications were performed.

MODIFICATIONS

A number of modifications were made on the boring mill, including removing all sections that protruded above the level of the rotating table. The two sides were cut off. The top sections were removed, including the cutting heads and the associated motors used for machining purposes. The oiling scheme was redone to feed the top of the gear box with oil. An oil return line and an oil drain line were plumbed in. The boring mill was color-coded by painting with different colors.

FEATURES

HIRB is a variable speed machine with a mechanical drive system equipped with four gears. Gears are shifted before starting the centrifuge. The gears drive a tapered roller bearing, which sits in a pool of oil. An oil pump, located in the gear box, pumps Mobil DTE-25 hydraulic oil throughout the gear box and the tapered roller bearing to provide lubrication during centrifuge operation. A 1.575 m diameter slotted steel table rests on the tapered roller bearing and is the main rotating part of the centrifuge. HIRB has a maximum rotation rate of 90 rpm. The mechanical drive system is connected to a DC motor via a coupling. A DC6 regenerative control system is used to control the motor and the centrifuge. The DC motor is a 30 hp, 230V, 125 A Allis-Chalmers motor. A feedback tachometer is connected to the motor shaft via a pulley-belt arrangement. The control system reads the output signal from the feedback tachometer and controls the motor speed accurately within 1%. For safety purposes, a steel chain link fence was constructed around the entire centrifuge. The fence is draped with two layers of thick felt cloth. Figures 2 and 3 are photographs showing the front and top views of the centrifuge. Table 1 lists some salient features of HIRB.

HIRB has several advantages over other commercial centrifuges. Since it is not used as a boring mill for machining purposes and its only purpose here is to spin an experimental package, the centrifuge operates under an almost no-load condition if we neglect air drag and some bearing friction. This centrifuge can run continuously for extended periods with no problems. The rotating table can accommodate extremely large and heavy loads. This centrifuge is also designed to take somewhat off-balance loads. It has a constant rotation rate and requires very little maintenance.

EQUIPMENT

Two steel I-beams are bolted to each other and to the rotating table. The lower I-beam is 20.32 cm high and 1.524 m long; its primary function is to raise the level of the second I-beam above the gear box on the back. The upper I-beam is 15.24 cm high and 3.048 m long and serves as the arm of the centrifuge. A steel arm mount is bolted to one end of the centrifuge arm. An aluminum swing bucket houses the furnace. This swing bucket is attached to the arm mount via two rod ends and a steel shaft. The steel shaft is held in place by a clevis pin and a cotter pin. As the centrifuge rotation rate increases, the swing bucket containing the furnace swings outward and aligns itself with the resultant acceleration (centrifugal acceleration plus earth's gravity). At the

Table 1. Salient features of HIRB

Centrifuge	Diameter of rotating table	1.575 m
	Max. rotation rate of table	90 rpm
	Motor	30 hp, 230 VDC, 125 A
	Control system	DC-6 Non-regenerative with tachometer feedback control
	Accuracy	Speed regulation of <1%
	Centrifuge drive	Mechanical drive via gears
	Radius of arm	1.524 m
	Max. acceleration level	$\approx$ 13.8 g at r = 1.524 m
	Radial position of the hinge	1.651 m
	Weight of package	20 kg
	Dimensions of package	30.48 $\times$ 30.48 $\times$ 25.4 cm
Slip rings	Power slip rings	110V, 30 A, 3 rings
	Instrumentation slip rings	28 V, 3 A, 22 rings
Instrumentation	Triaxial acclerometer and power supply	$\pm$25 g, 0 - 300 Hz
	Silicon controlled rectifier	4 - 20 mA signal
	Data acquisition system	20 channel thermocouple board, 8 channel analog input board, 8 channel analog output board, 80486 33 MHz computer, Custom written control software

Figure 2. Front View of HIRB.

Figure 3. Top view of HIRB.

other end of the arm of the centrifuge, 1.27 cm thick steel plates act as counterweights to balance the load acting on the furnace end of the beam.

SLIP RINGS

A custom-built slip ring assembly was purchased from Fabricast Inc. of South El Monte, CA. The slip ring assembly contains 3 power slip rings and 22 instrumentation slip rings, and is rated for a maximum rotation rate of 90 rpm. The power slip rings are made of copper and rated for 110V, 30A. The instrumentation slip rings are made of coin silver and are rated for 28 V, 3A with a low dynamic resistance of 10 milliohms. Silver graphite brushes are used. Slip rings are enclosed in a one-piece, cadmium-plated steel sleeve rotor. The whole assembly is housed in an anodized aluminum housing. Rotor leads are 3.048 m long and made of 20 gage military specification wire(MIL-W-16878 Type "E"). This wire is stranded, silver coated, and shielded and jacketed with a Teflon sheath. The slip ring assembly is mounted on a rotor slip ring mount made of steel. This mount enables us to bolt down the slip ring assembly to the arm of the centrifuge and to feed the rotor leads out through the bottom of the slip ring assembly to the arm. The stator is held stationary by a stator slip ring mount made of 7.62 cm diameter polyvinyl chloride pipe. The PVC pipe is attached to a steel unistrut, which is bolted to the fence and supported by six steel cables.

FURNACE

A one-zone gradient freeze furnace was constructed in the crystal growth laboratory at Clarkson University. A helically wound 18 gage Kanthal A-1 wire is the heating wider spacing along the bottom, so as to produce a temperature gradient down the furnace length. The furnace is 27.94 cm long, 16.19 cm diameter and has a core diameter of 3.18 cm. The core is made of Fibercraft moistened formable insulation obtained from Thermcraft Incorporated. The core is wrapped with layers of fiber blanket insulation obtained from Thermcraft Incorporated. A quartz liner with dimensions of 22 mm x 17 mm and 25.4 cm long is used to hold the ampoule in the furnace. The liner is positioned in the center of the furnace by two ceramic inserts, one on the top and one on the bottom. The furnace has a cylindrical outer shell of 16 gage aluminum and two end caps of 4 gage aluminum. The end caps are held in place by 12 right angle brackets via countersunk screws. Electrical and thermocouple contacts are made through the side wall onto a strip of insulating G-10 glass epoxy. Three K-type thermocouples are embedded in the furnace side wall. Two more thermocouples are placed along the center of the furnace, one above the ampoule and one below the ampoule.

Figure 4 is a photograph of the furnace mounted in the swing bucket on the centrifuge. Figure 5 shows the temperature profile inside the empty furnace using a thermocouple positioned along the center of the furnace.

INSTRUMENTATION

A triaxial accelerometer was purchased from Entran Devices in Fairfield, NJ. The miniature triaxial accelerometer is mounted on the swing bucket to measure the acceleration along all three axes. The accelerometer is a 1.27 cm cube and is mounted on the swing bucket via a tapped hole. This accelerometer permits simultaneous acceleration, vibration and shock measurement in three perpendicular axes. It is a piezoresistive accelerometer that employs a fully active semiconductor Wheatstone bridge. A high output of 5 mV/g enables the accelerometer to drive data monitoring systems directly, without amplification or costly signal conditioning. The semiconductor circuitry is fully compensated for temperature changes in the environment and possesses excellent thermal characteristics. This triaxial accelerometer has a range of ±25 g, a useful frequency range of 0 - 300 Hz and an operating temperature range of 233 K to 394 K. An excitation voltage of 15 VDC is supplied by a precision adjustable power supply. The power supply is mounted on a control panel at the center of the rotating table and is powered by a 110 VAC supply.

DATA ACQUISITION AND CONTROL

The gradient freeze furnace is interfaced to an 80486 IBM-compatible computer and a Keithley Metrabyte data acquisition system. An user-friendly data acquisition and PID furnace control program was written, debugged and implemented. Analog signals are sent between the computer and the furnace via slip rings for furnace temperature control and data acquisition. A Eurotherm silicon-controlled rectifier is mounted on a control panel at the axis of rotation of the centrifuge table. From the computer, the rectifier receives a 4-20 mA signal that depends on the set point. The rectifier provides the corresponding power output to the furnace. The data acquisition system collects data from five thermocouples mounted in the furnace and from the triaxial accelerometer. All these data are displayed on the computer monitor in real-time along with the temperature-time curve of the growth process.

Figure 4. Gradient freeze furnace mounted in the swing bucket on the centrifuge for crystal growth experiments.

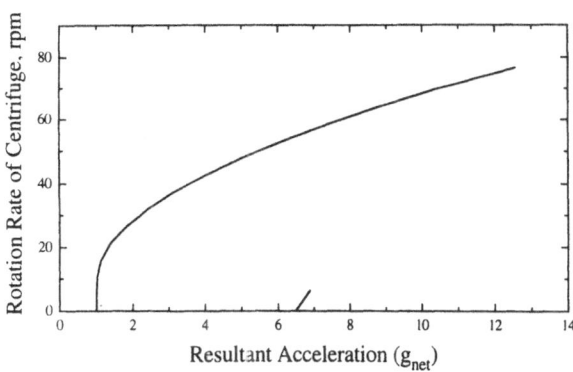

Figure 5. Temperature profile in the empty gradient freeze furnace measured along the center axis.

CALIBRATION

An accurate calibration of the acceleration at the end of the arm of the centrifuge was performed. We believe this is the most accurate measurement of the g level experienced by an ampoule during crystal growth experiments. The centrifuge swing bucket and furnace assembly were designed and constructed such that the center of mass acts along the longitudinal axis of the furnace. Only if the center of mass is on the furnace axis would the ampoule be aligned with the resultant g vector during centrifugation. A calibration jig was designed and constructed to mount the triaxial accelerometer inside the furnace. The accelerometer was mounted at the same spot where the center of the ampoule would be during crystal growth experiments. The centrifuge was spun at different rotation rates and the resultant acceleration was measured by the accelerometer. Since our furnace was well aligned, the g levels on two axes were zero, while the third axis represented the resultant g level. A calibration curve for the entire range of rotation rates of the centrifuge is shown in figure 6.

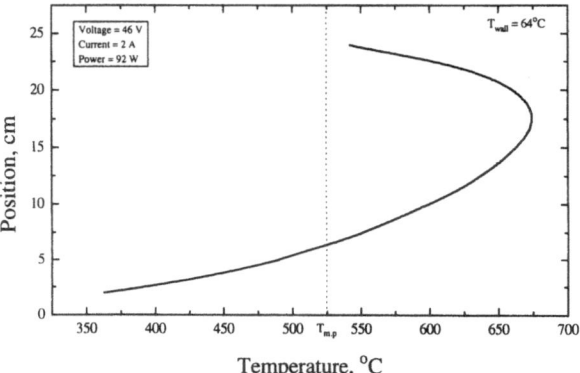

Figure 6. Calibration curve for HIRB.

CONCLUSIONS

The centrifuge "HIRB" at Clarkson University's International Center for Gravity Materials Science and Applications is a unique facility. It has wide-ranging possibilities for experiments in different areas of materials processing. We expect this centrifuge to provide improved insight into the effect of high gravity on materials processing.

ACKNOWLEDGEMENTS

This work was supported by Clarkson University and by New York State via the Center for Advanced Materials Processing. We thank the Physical Plant and Maintenance Department at Clarkson University for help in setting up the centrifuge. Furthermore we would like to thank all the people whose help we could not have done without during the construction of this centrifuge facility.

REFERENCES

1. L.L. Regel, "Materials Processing in Space: Theory, Experiments, Technology," Consultants Bureau, Division of Plenum Publishing Corporation, New York, (1990).
2. H. Rodot, L.L. Regel, G.V. Sarafanov, M. Hamidi, I.V. Videskii and A.M. Turtchaninov, *J. Crystal Growth* 79:77 (1986)
3. H. Rodot, L.L. Regel and A.M. Turtchaninov, *J. Crystal Growth* 280:104 (1990).
4. R. Derebail, W. R. Wilcox and L. L. Regel, *J. Crystal Growth* 119:98 (1992).
5. R. Derebail, W. R. Wilcox and L. L. Regel, *J. Spacecraft and Rockets* 30:202, No. 2 (1993).

ESTABLISHMENT OF THE NEW
C-CORE CENTRIFUGE CENTRE

M.J. Paulin, R. Phillips, J.I. Clark, R. Meaney, D. Millan and K. Tuff

C-CORE - Centre for Cold Ocean Resources Engineering
Memorial University of Newfoundland
St. John's, Newfoundland
A1B 3X5 Canada

ABSTRACT

The C-CORE Centrifuge Centre was completed and commissioned in May, 1993. Located on the campus of Memorial University of Newfoundland in St. John's, Newfoundland, Canada, this new facility physically links C-CORE (Centre for Cold Ocean Resources Engineering) and the Faculty of Engineering and Applied Science, both of which will have close working relationships with the Centrifuge Centre. This centre is a new two-story building which was designed especially for and dedicated to centrifuge modelling. At the heart of the facility is an Acutronic 680-2 centrifuge similar to the one at the Laboratoire Centrale des Ponts et Chaussées in Nantes, France. The machine is 5.5 metres in radius and can carry a payload of 2200 kg to 100 g or a payload of 650 kg to 200 g. Models up to 1.1 by 1.4 m in plan and 1.1 m in height can be accommodated on the St. John's machine. The machine is the largest of its kind in Canada. A custom-built 7 kW refrigeration unit supplies coolant down to a temperature of -35 °C to the experimental package through two of the centrifuge's six rotary joints. Currently, this is the only machine in North America with this capability. The other rotary joints are used to supply 200 bars of hydraulic pressure and 10 bars of pneumatic pressure. The electrical slip ring stack comprises: 64# - 1 amp signal lines; 8# - 15 amp power lines; a 40 kW, 3-phase, 80 amp, power supply; and 6 coaxial channels.

INTRODUCTION

C-CORE (Centre for Cold Ocean Resources Engineering) is an independently funded research institute located at Memorial University of Newfoundland in St. John's, Newfoundland, Canada. Funded by industry and government, the Centre has traditionally undertaken research, development and technology transfer that contribute to the safe and productive use of Canada's ocean resources. This research was conducted by three main groups at C-CORE: Ice Engineering, Seabed Geotechnics, and Remote Sensing.

In 1992, a fourth group was formed at C-CORE, a Geotechnical Engineering Group. The creation of this group reflected C-CORE's growing expertise in geotechnical engineering and the culmination of a major initiative to acquire a technically advanced centrifuge modelling center for St. John's. Although the C-CORE centrifuge is the first large centrifuge in Canada, the technology is well developed and has been extensively applied to the solution of engineering problems elsewhere in the world. Indeed, many Canadian companies have made use of large centrifuge testing facilities in the United States, United Kingdom, and France.

Centrifuge modelling can be used to examine problems in which self-weight or ambient acceleration plays a role. Traditionally, centrifuge modelling has been used to investigate geotechnical problems (see, for example, Schofield[1]) including soil statics, soil dynamics, cold regions studies, and hazardous waste disposal. Other areas of application are earth sciences (e.g. structure formation and reservoir engineering) and material processing research. New applications of centrifuge modelling are being developed as new engineering or scientific problems arise. Such new areas include: crystal growth, ice research, aeronautics testing, and component testing.

THE C-CORE CENTRIFUGE CENTRE

The C-CORE Centrifuge Centre is located between the Captain Robert A. Bartlett building and the S.J. Carew building on the campus of Memorial University of Newfoundland. The Centrifuge Centre was constructed and equipped through funding from the Canada-Newfoundland Offshore Development Fund, the Technology Outreach Program of Industry, Science and Technology Canada and the Natural Sciences and Engineering Research Council Canada.

The centre comprises a two-story building, containing laboratories and workshops with offices upstairs and a containment structure housing an Acutronic 680-2 centrifuge. The containment structure has three levels. The upper level provides a stiff ceiling to the main centrifuge chamber to resist the aerodynamic excitation imposed by the centrifuge in rotation. The upper level also houses the electrical slip ring capsule and associated interfaces. The intermediate level is the main centrifuge chamber, which is accessible by forklift from the main building. The main chamber is 13.5 m in diameter and 4.2 m high. The 300 mm thick reinforced concrete chamber wall is aerodynamically clean inside and retains a rockfill safety berm outside. The lower level is underground and contains the centrifuge drive unit with associated controllers and a refrigeration unit. The two-story building includes sample preparation and investigation areas, an x-ray bay, mechanical and electrical workshops, a coldroom, data processing areas and offices, including areas for visiting researchers or clients. The building also has access to Memorial University's computer capabilities. The building is shown in Fig. 1.

THE ACUTRONIC 680-2 CENTRIFUGE

This Acutronic 680-2 centrifuge, shown in Fig. 2, is capable of testing models to 200 g and has a radius of 5.5 m to the surface of the swinging platform. The test package is usually at a nominal working radius of 5 m. At the centrifuge maximum rotational speed of about 189 rpm, the acceleration at the package is about 200 g. The C-CORE centrifuge has a maximum payload capacity of 100 g x 2.2 t = 220 g-tonnes capacity at 5 m radius. This capacity reduces to 130 g-tonnes at 200 g due to the increased self-weight of the platform.

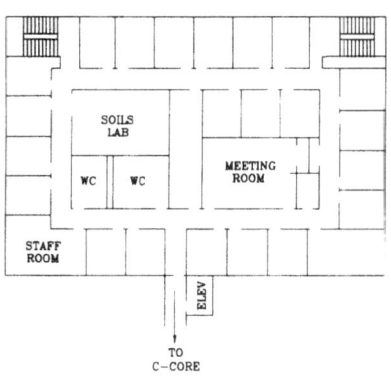

SECOND FLOOR

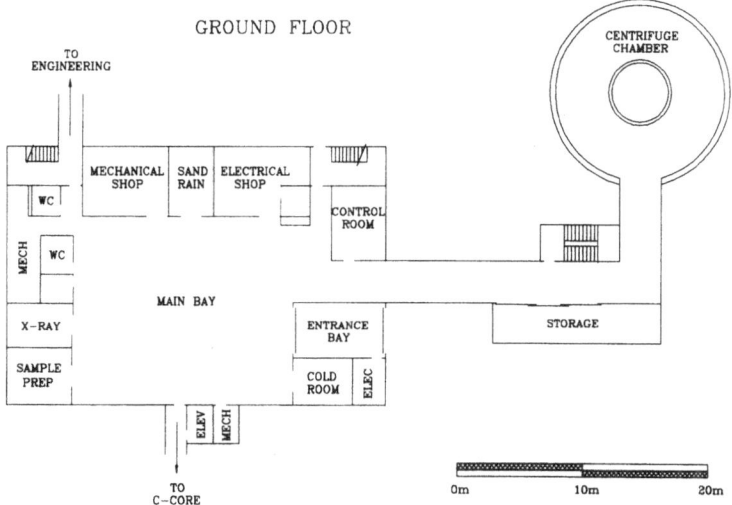

GROUND FLOOR

0m   10m   20m

Figure 1. C-CORE Centrifuge Centre

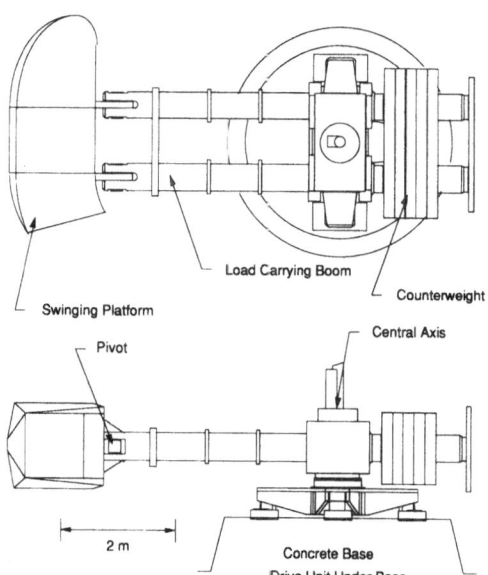

Load Carrying Boom

Counterweight

Swinging Platform

Pivot

Central Axis

2 m

Concrete Base

Drive Unit Under Base

Figure 2. Acutronic 680-2 centrifuge

The capacity and specifications of the Acutronic 680-2 are presented in Fig. 3. The maximum payload size is 1.1 m high by 1.4 m deep and 1.1 m wide. This centrifuge is similar to the one operating at the Laboratoire Centrale des Ponts et Chaussées in Nantes, France.

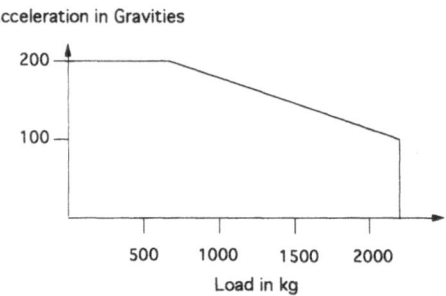

Figure 3. Capacity of the Acutronic 680-2

The centrifuge arm consists of two parallel steel tubes held apart by a central drive box and spacers, as shown in Fig. 4. The swinging platform is suspended on bushings from the ends of the steel tubes. The swinging platform is covered by an aerodynamic shroud to reduce drag. The platform and the payload are balanced by a 20.2 tonne mass counterweight. The position of this counterweight is adjusted by driving a series of gearwheels along screwthreads on the outside of the parallel steel tubes using an electric motor.

Figure 4. C-CORE Centrifuge

The centrifuge arm rotates on a set of tapered roller bearings inside the central drive box and mounted on a stationary shaft. This shaft is attached to the centrifuge containment through a four branch star support suspended on four springs. Each of the four springs is strain-gauged to sense imbalance within the centrifuge arm to within 10 kN.

The centrifuge drive unit comprises a 450 kW AC variable speed motor and a 9:1 gear reducer. The variable speed motor in energized through two 250 kW invertors connected in parallel. Precision couplings and a hollow vertical drive shaft connect the hollow output shaft of the gear reducer to the central drive box. Two rotary joints are attached beneath the gearbox output shaft, which contain 6 passages and is described below.

The power consumption is due mainly to aerodynamic drag within the centrifuge chamber. The centrifuge and the chamber are cooled by forced air ventilation. Air is drawn into the chamber through a ceiling vent around the central axis of the centrifuge. Air is drawn out of the chamber through a floor vent by an exhaust fan located in the lower level.

CENTRIFUGE SERVICES

Rotary Joints

The Acutronic 680-2 is equipped with two rotary joints, which permit fluids to flow through the central axis of the machine to the platform. These rotary joints contain a total of six passages; two are designed to accept high pressure hydraulic fluid; two are dedicated to the refrigeration unit (glycol refrigerant); and the remaining two can carry either air or water. The rated capacities of the fluid passages are as follows:

High pressure joints: working pressure: 200 bar (20 MPa)
working temp: 10 to 50 °C

Cold fluid joint: working pressure: 20 bar (2 MPa)
working temp: -30 to 50 °C

Air/water joint: working pressure:
air: 7 bar (725 kPa)
water: 20 bar (2 MPa)
working temp: 10 to 50 °C

Refrigeration Unit

To enable cooling of the experimental package, the C-CORE Centrifuge Centre has a refrigeration system. The refrigeration system is designed to deliver 7 kW of cooling to the platform. Cooling of the package is accomplished by pumping a cold fluid through the rotary joint to the end of the boom. The refrigeration unit can deliver 10 ℓ/min of glycol refrigerant with temperatures reaching -30 °C. At the platform, a series of fluid-to-air or fluid-to-fluid heat exchangers provide cooling of the package. Temperature control is accomplished by varying the volume of refrigerated fluid permitted to pass through these heat exchangers. This system is depicted schematically in Fig. 5.

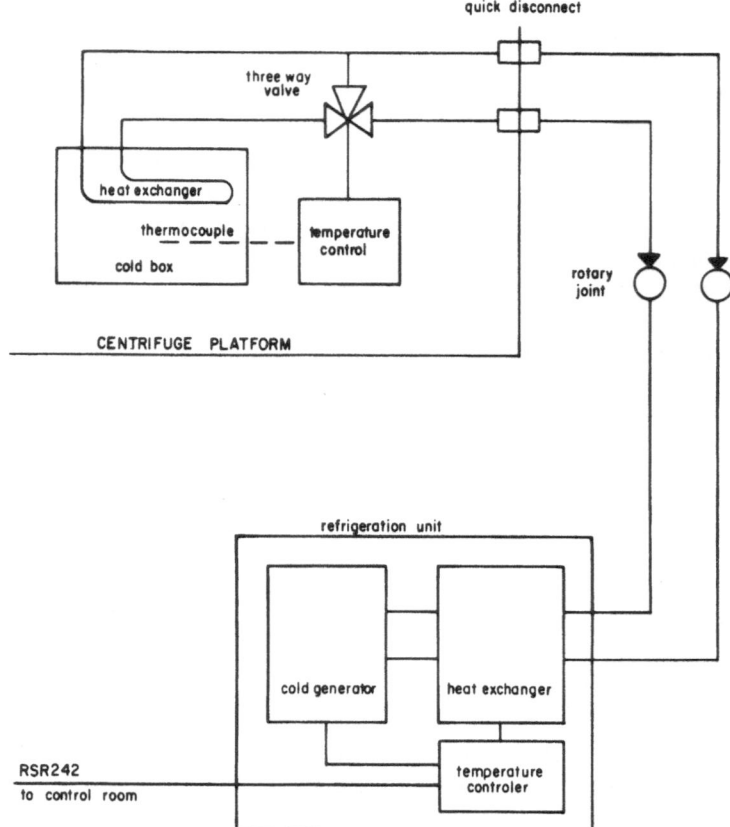

Figure 5. C-CORE Centrifuge refrigeration system

Ancillary Equipment

Ancillary equipment are currently being developed to support operations within the Centrifuge Centre. These essential items include strongboxes, consolidometers, an in-flight cone penetrometer, and a refrigerated strongbox. The design of these devices is complicated by the level of stresses experienced in the centrifuge and by restrictions on size and weight. Additionally, mechanical or electrical systems must be controlled remotely.

Primarily, the strongboxes or tubs act as containment structures for all experiments to be conducted on the centrifuge, and provide bases to which experimental apparatus can be secured during flight. The existing strongboxes have an internal working area of 1.18 m x 0.94 m and a depth of 0.4 m. However, additional tubs can be constructed as the need arises to suit individual experimental needs.

The cold box is a custom strongbox used to house experiments requiring temperature control. The design temperature range for this cold box has been specified as -25 °C to +10 °C. Temperatures will be controlled remotely by directing the refrigerated fluid either into or away from the cold box heat exchangers.

A consolidometer is a large hydraulic press used in sample preparation. The consolidometers are designed to apply consolidation pressures of up to 400 kPa onto samples contained in the existing strongbox. The load is delivered by a 200 mm diameter hydraulic cylinder with a working pressure of 17 MPa, which equates to approximately 50 tonnes of static thrust.

In-flight mechanical properties of test samples are required during centrifuge modelling. One means of acquiring these data is through the use of a cone penetrometer. This device consists of an instrumented shaft and several linear actuators. The instrumented shaft is capable of measuring tip and shaft resistance. The cone penetrometer will utilize two linear actuators. One actuator will drive the instrumented shaft into the sample. This actuator is designed for a peak load of 10 kN and a maximum penetration rate of 20 mm/sec. The second actuator is used to position the cone within the test package.

DATA ACQUISITION AND ELECTRICAL POWER SUB-SYSTEMS

Some of the more important design criteria for the data acquisition and electrical power sub-systems include:

- The highest degree of data integrity, resolution, and accuracy possible.
- A system that can be maintained and supported in a *high-throughput* industrial environment.
- Extreme flexibility in supporting instrumentation and control functions.
- A slipring capsule providing the following:
    - 64 x 1 amp signal rings
    - Common 8 x 15 amp electrical power/control rings
    - 3 x 50 ohm coaxial rings
    - 3 x 75 ohm coaxial rings
    - Full 3-phase, 5-wire, 380 V, 80 amp, 60 Hz power service.

The data acquisition and electrical power sub-system in place at the Centrifuge Centre is depicted in Fig. 6. Currently, the system is limited to a maximum of 64 channels of data acquisition. In most experiments this number will be less because some of the signal lines will have to be used for control and communication.

The design utilizes a high quality custom-designed signal conditioning (S/C) sub-system mounted on the strongbox. The individual S/C modules are dual channel printed circuit cards, mounted in a 12 card chassis. Each chassis provides on-board regulated excitation supplies for the attached instrumentation. Each individual channel/instrument is fed via a six pin Circular-Mil connector; the card/connector is configurable on the S/C card for a wide variety of instruments. The bulk power for the S/C cards and instrumentation is fed from a high quality power supply mounted on the drive box, via a bulk power umbilical. The power supplies are fed from a 220 V single phase, 3-wire connection from the main 380 V, 5-wire, 3-phase supply.

The signals from each chassis are fed back to a patch panel on the drive box where they are redistributed to suit the 2 x 32 channel arrangement of the slip rings. On the far side of the sliprings the signals run into a shielded cabinet and are attached to a 64 channel multiplexer, which then feeds a PC based Analogic HSDAS-16 (16 bit A/D convertor). The SlipRing Room data acquisition PC is at present running SNAPSHOT data acquisition software. This PC is connected via a thin wire Ethernet to the Control Room Data Acquisition PC, and logs all its data to the CRDAS PC's Magneto-Optical drive. The various coaxial rings are rated to carry data-communications and high bandwidth analog signals.

Most of the lab/support equipment for maintaining, calibrating and repairing the elements of the data acquisition/power systems is in place.

In summary, the present data acquisition system is an intermediate step on the way to a state-of-the-art system. For the present, this system fulfills the data acquisition needs of the centre but can be easily upgraded as the need arises.

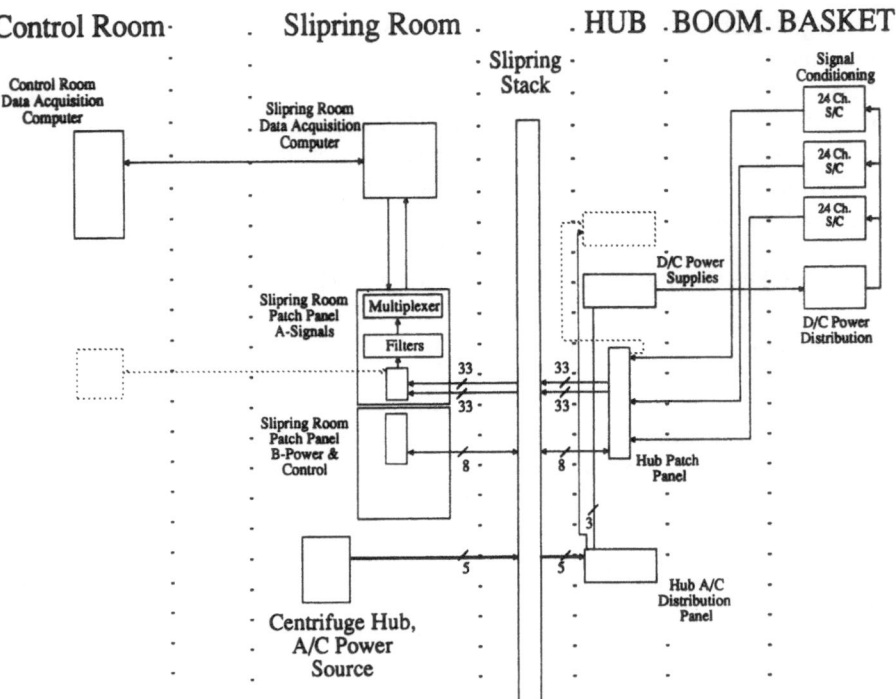

Figure 6. Data acquisition and electrical power sub-systems

REFERENCE

1. A.N. Schofield, "Twentieth Rankine Lecture: Cambridge Geotechnical Centrifuge Operations," *Geotechnique*, 30:227 (1980).

INDEX